STUDENT UNIT GUIDE

WJEC A2 Biology Unit BY5
Environment, Genetics and Evolution

Dan Foulder

Philip Allan, an imprint of Hodder Education, an Hachette UK company, Market Place, Deddington, Oxfordshire OX15 0SE

Orders
Bookpoint Ltd, 130 Milton Park, Abingdon, Oxfordshire OX14 4SB
tel: 01235 827827
fax: 01235 400401
e-mail: education@bookpoint.co.uk
Lines are open 9.00 a.m.–5.00 p.m., Monday to Saturday, with a 24-hour message answering service. You can also order through the Philip Allan website: www.philipallan.co.uk

ISBN 978-1-4441-8300-9

First printed 2013
Impression number 5 4 3
Year 2015

Cover photo: Fotolia

Typeset by Integra Software Services Pvt. Ltd., Pondicherry, India

Printed in India

Hachette UK's policy is to use papers that are natural, renewable and recyclable products and made from wood grown in sustainable forests. The logging and manufacturing processes are expected to conform to the environmental regulations of the country of origin.

This material has been endorsed by WJEC and offers high quality support for the delivery of WJEC qualifications. While this material has been through a WJEC quality assurance process, all responsibility for the content remains with the publisher.

P2198

Contents

Content Guidance

Questions & Answers

Getting the most from this book

Examiner tips

Advice from the examiner on key points in the text to help you learn and recall unit content, avoid pitfalls, and polish your exam technique in order to boost your grade.

Knowledge check

Rapid-fire questions throughout the Content Guidance section to check your understanding.

Knowledge check answers

1 Turn to the back of the book for the Knowledge check answers.

Summary

Summaries

- Each core topic is rounded off by a bullet-list summary for quick-check reference of what you need to know.

Questions & Answers

Exam-style questions

Examiner comments on the questions
Tips on what you need to do to gain full marks, indicated by the icon ⓔ.

Sample student answers
Practise the questions, then look at the student answers that follow each set of questions.

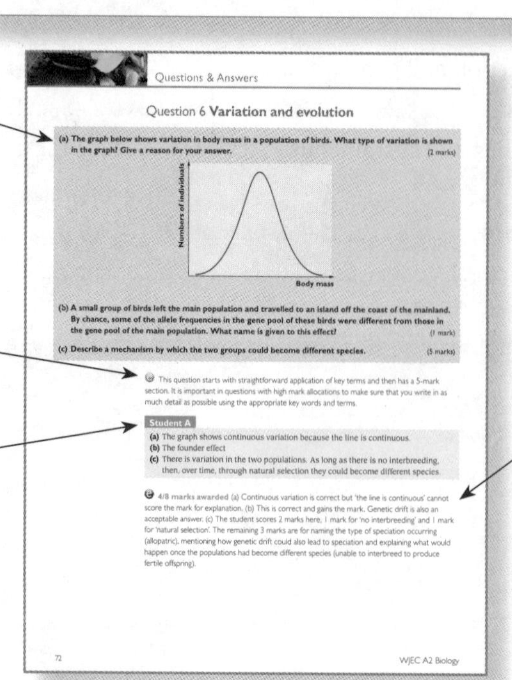

Questions & Answers

Question 6 **Variation and evolution**

(a) The graph below shows variation in body mass in a population of birds. What type of variation is shown in the graph? Give a reason for your answer. (2 marks)

(b) A small group of birds left the main population and travelled to an island off the coast of the mainland. By chance, some of the allele frequencies in the gene pool of these birds were different from those in the gene pool of the main population. What name is given to this effect? (1 mark)

(c) Describe a mechanism by which the two groups could become different species. (5 marks)

ⓔ This question starts with straightforward application of key terms and then has a 5-mark section. It is important in questions with high mark allocations to make sure that you write in as much detail as possible using the appropriate key words and terms.

Student A

(a) The graph shows continuous variation because the line is continuous.
(b) The founder effect
(c) There is variation in the two populations. As long as there is no interbreeding, then, over time, through natural selection they could become different species.

ⓔ **4/8 marks awarded** (a) Continuous variation is correct but 'the line is continuous' cannot score the mark for explanation. (b) This is correct and gains the mark. Genetic drift is also an acceptable answer. (c) The student scores 2 marks here, 1 mark for 'no interbreeding' and 1 mark for 'natural selection'. The remaining 3 marks are for naming the type of speciation occurring (allopatric), mentioning how genetic drift could also lead to speciation and explaining what would happen once the populations had become different species (unable to interbreed to produce fertile offspring).

72 WJEC A2 Biology

Examiner commentary on sample student answers
Find out how many marks each answer would be awarded in the exam and then read the examiner comments (preceded by the icon ⓔ) following each student answer.

About this book

The aim of this book is to help you to prepare for the WJEC A2 Biology Unit BY5 examination.

The **Content Guidance** section contains everything you need to learn to cover the specification content of BY5. This is a varied unit with a large content volume. Across all the topics there are a number of processes with stages to learn — for example, oogenesis, DNA fingerprinting and succession. It is important that you are clear on all the stages of these processes and can describe them fully. It is also important you can apply the content to new situations and to data that you are given. For example, in the past the BY5 examination has included questions on the distribution of jackal populations and the effects of burning on upland heather. It would be impossible to revise all the areas you could be asked about, but if you revise the content of the specification thoroughly and think the question through carefully then you should be able to apply your knowledge. You should also be comfortable with carrying out the different types of genetic crosses included in the specification — a large number of marks are allocated to such questions.

A small number of marks on the BY5 exam are synoptic. This means they are based on AS content, which will be part of a wider BY5 question. Therefore, while studying this module you should try to think about how AS content may fit into the BY5 topics.

The **Questions and Answers** section contains questions on each of the topic areas in the specification. They are written in the same style as the questions in the BY5 exam so they will give you an idea of the sort of thing you will be asked to do in the exam. After each question there are answers by two different students followed by examiner's comments on what they have written. These are important because they give you an insight into the responses the examiners are looking for in the exam. They also highlight some of the common mistakes students make.

Revision techniques

It is important to develop effective revision and study techniques. The key to effective revision is to make it *active*. Most people cannot revise effectively by just reading through notes or a book. In order to learn you have got to do something with the information. Below are some examples of active revision techniques — not all of them will work for everybody so it is important that you try them and find out which ones work for you.

Consolidate your notes

This means taking information from your notes and this book and presenting it in a different form. This can be as simple as just writing out the key points of a particular topic. One effective consolidation technique involves taking some information and turning it into a table or diagram. More creative consolidation techniques include the use of mind maps or flash cards.

The key to these techniques is that you will be actively thinking about the information you are revising. This increases the chance of you remembering it and also allows you to see links between different topic areas. Developing this kind of deep, holistic understanding of the content of BY5 is the key to getting top marks in the exam.

Complete practice exam questions

Completing practice exam questions is a crucial part of your revision. It allows you to practise applying the knowledge you have gained to the exam questions and see if your revision is working. A useful strategy is to complete questions on a topic you have not yet revised fully. This will show you which areas of the topic you know already and which areas you need to work on. You can then revise the topic and go back and complete the question again to check that you have successfully plugged the gaps in your knowledge.

Use technology

There are many creative ways to use technology to help you revise. For example, you can make slideshows of key points, shoot short videos or record podcasts. The advantage of doing this is that by creating the resource you are thinking about a particular topic in detail. This will help you to remember it and improve your understanding. You will also have the finished product, which you can revisit closer to the exam. You could even pass on what you have made to your friends to help them.

Content Guidance

DNA replication and protein synthesis

The function of DNA is to carry the **genetic code** for an organism. The genetic code is the code for protein synthesis. It is therefore fundamental to the life of any organism.

Genetic stability is also important to organisms. It relies on the genetic code being passed on to daughter cells without changes or errors. Mutations are rarely beneficial. They can lead to non-functional cells being produced or, in the worst cases, may even threaten the survival of the whole organism — for example a mutation that leads to the activation of an oncogene and the onset of cancer. To ensure that genetic stability is preserved it is therefore important that DNA is replicated accurately. The structure of DNA is shown in Figure 1.

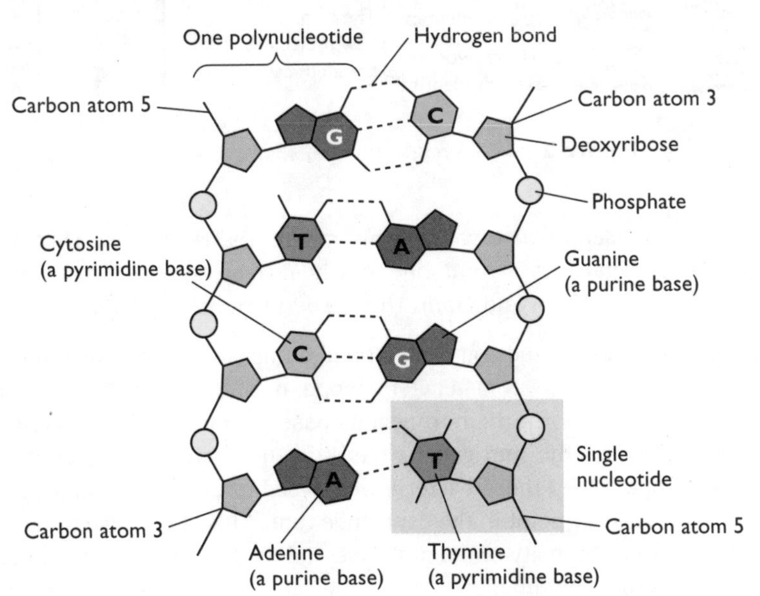

Figure 1 The molecular structure of DNA, showing the two antiparallel polynucleotide strands

Examiner tip
You should revise the structure of DNA from BY1, with particular emphasis on how its structure allows exact replication of the molecule, i.e. that the bases always pair in a complementary way, adenine with thymine and cytosine with guanine.

The process of DNA replication is shown in Figure 2 and is outlined below:

- The DNA unwinds and the hydrogen bonds holding the two strands of DNA together break; this is sometimes referred to as unzipping.
- The enzyme DNA polymerase catalyses the addition of free DNA nucleotides to form two new complementary strands (adenine bonding with thymine and cytosine bonding with guanine) using both original strands as templates.
- Hydrogen bonds then form between each pair of complementary DNA strands, (one original strand, one new strand). This produces two molecules of DNA that are identical to each other and to the original DNA molecule.

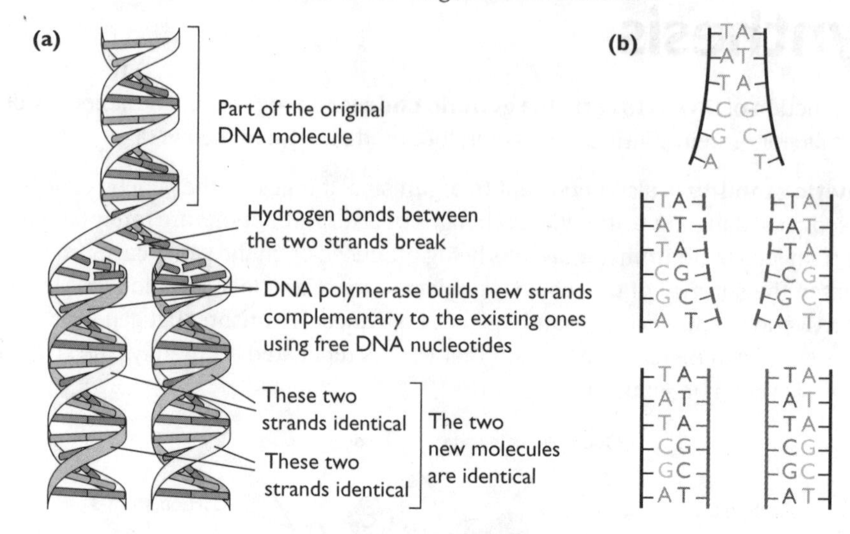

Figure 2 (a) DNA replication; (b) semi-conservative replication of DNA

Examiner tip

Questions on Meselson and Stahl's experiment occur frequently. It is a complex experiment, so make sure that you learn it thoroughly.

Knowledge check 1

Why is DNA replication said to be semi-conservative?

DNA replication is semi-conservative. This means that each new DNA molecule contains one original strand and one newly formed strand. This was shown experimentally by Meselson and Stahl. Their experiment is detailed below.

Meselson and Stahl cultivated some bacteria in a flask of nutrient medium in which the source of nitrogen was ^{15}N (a heavier isotope than ^{14}N). The bacteria took up the nitrogen and used it to form the nitrogenous bases of nucleotides. Therefore, after several generations all the nitrogenous bases in the bacterial DNA contained the heavier ^{15}N isotope. When these bacteria were lysed (broken open) and centrifuged the DNA settled at a low point in the centrifuge tube. This is because the position in the tube relates to the density of the molecule and as the DNA contained ^{15}N, it was denser than DNA that contained ^{14}N.

They then allowed the bacteria to divide *once* in a medium that contained ^{14}N (the lighter isotope). The bacteria were lysed and centrifuged. The DNA settled at a midway point in the tube, indicating that it was less dense than the DNA in the first generation. This is because all the bacterial cells now contained 'hybrid' DNA — DNA that contains one original ^{15}N-containing strand and a newly formed strand that also contained ^{14}N in its nitrogenous bases.

This result proved that DNA replication was not conservative. The theory of conservative replication suggested that the original double-stranded DNA molecule remained and a totally new double-stranded molecule was produced. By forming hybrid DNA Meselson and Stahl proved this to be incorrect.

However this result could be explained by semi-conservative replication (one original strand and one new strand) or dispersive replication (both strands consisting of sections of new and old DNA). Meselson and Stahl allowed the bacteria to divide a second time in the growth medium containing ^{14}N. When they extracted the DNA and centrifuged it, a hybrid band again formed at a midway point up the tube and another band of less dense DNA containing only ^{14}N occurred further up the tube. This result proved that the DNA is replicated semi-conservatively. Meselson and Stahl's investigation is illustrated in Figure 3.

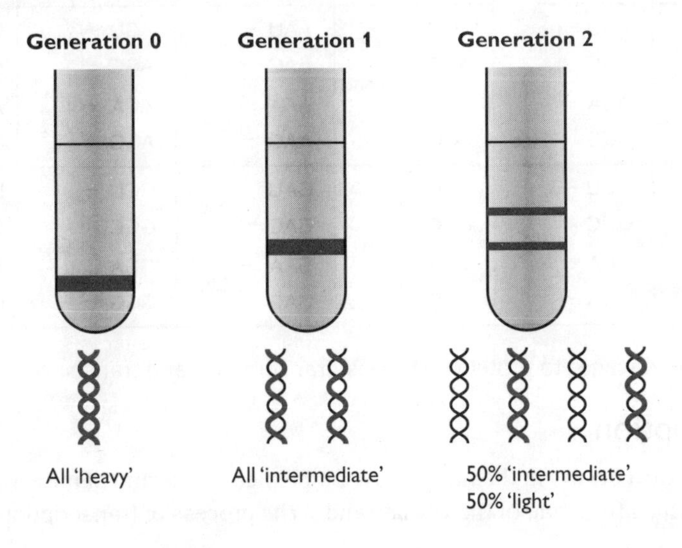

Figure 3 Meselson and Stahl's experiment

Protein synthesis

Proteins are of vital importance in cells; they are made up of one or more polypeptide chains. The code for the structure of the polypeptide is carried in the DNA of the cell. During protein synthesis this code is 'read' and the amino acids are arranged in the correct sequence to form the polypeptide chain. The sequence of nucleotide bases coding for one polypeptide is a gene. Genes are made up of triplets of bases (codons). Each triplet of bases codes for one amino acid. There are only 20 amino acids and 64 possible triplets of bases (the number of possible combinations of the four different nucleotide bases in DNA in groups of three), and almost all amino acids are coded for by more than one triplet of bases. There are also stop and start codes that are used to control protein synthesis. During the first stage of protein synthesis (transcription) the genetic code on a section of DNA is transcribed to form a strand of messenger RNA (mRNA). The mRNA codons and the amino acids they code for are shown in Table 1.

Knowledge check 2

What is the section of DNA that codes for one polypeptide called?

Table 1 RNA codons

First position (5′ end)	Second position: U		C		A		G		Third position (3′ end)
U	UUU	Phe	UCU	Ser	UAU	Tyr	UGU	Cys	U
	UUC	Phe	UCC	Ser	UAC	Tyr	UGC	Cys	C
	UUA	Leu	UCA	Ser	UAA	stop	UGA	stop	A
	UUG	Leu	UCG	Ser	UAG	stop	UGG	Trp	G
C	CUU	Leu	CCU	Pro	CAU	His	CGU	Arg	U
	CUC	Leu	CCC	Pro	CAC	His	CGC	Arg	C
	CUA	Leu	CCA	Pro	CAA	Gln	CGA	Arg	A
	CUG	Leu	CCG	Pro	CAG	Gln	CGG	Arg	G
A	AUU	Ile	ACU	Thr	AAU	Asn	AGU	Ser	U
	AUC	Ile	ACC	Thr	AAC	Asn	AGC	Ser	C
	AUA	Ile	ACA	Thr	AAA	Lys	AGA	Arg	A
	AUG	Met	ACG	Thr	AAG	Lys	AGG	Arg	G
G	GUU	Val	GCU	Ala	GAU	Asp	GGU	Gly	U
	GUC	Val	GCC	Ala	GAC	Asp	GGC	Gly	C
	GUA	Val	GCA	Ala	GAA	Glu	GGA	Gly	A
	GUG	Val	GCG	Ala	GAG	Glu	GGG	Gly	G

Examiner tip
You do not have to learn the table of codons but you should be able to use it if given it in a question.

Knowledge check 3
Where does transcription occur?

Knowledge check 4
Name the enzyme that catalyses the formation of the mRNA strand during transcription.

There are two stages to protein synthesis: transcription and translation.

Transcription

Transcription occurs in the nucleus. A complementary strand of mRNA (messenger RNA) is formed from one of the DNA strands. The process of transcription is detailed below:

- A section of the DNA unwinds and the hydrogen bonds holding the two strands of the DNA break, (the DNA unzips).
- The enzyme RNA polymerase catalyses the addition of free RNA nucleotides to form a complementary strand of messenger RNA (mRNA). The complementary bases are:
  - cytosine on the DNA and guanine on the mRNA
  - guanine on the DNA and cytosine on the mRNA
  - thymine on the DNA and adenine on the mRNA
  - adenine on the DNA and *uracil* on the mRNA (thymine does not occur in RNA and uracil is the complementary base to adenine)
- When translation is complete, the hydrogen bonds between the two strands of DNA reform.
- The mRNA molecule then leaves the nucleus through the nuclear pore and travels to a ribosome in the cytoplasm.

The process of transcription is shown in Figure 4.

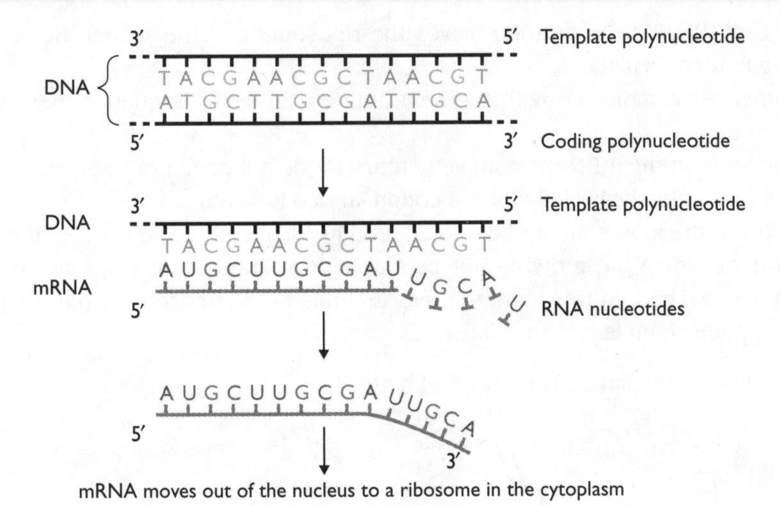

Figure 4 Transcription

Translation

The purpose of transcription is to transcribe the code for the polypeptide from DNA to mRNA so that it can be transported to the ribosome. In the next stage, **translation**, this code is used to form the polypeptide. In translation, transfer RNA (tRNA) is used to transfer amino acids to the ribosome.

The **anticodon** on the tRNA molecule determines the specific amino acid it transfers. The anticodon is complementary to the codon for that amino acid on mRNA. The specific amino acid joins to the amino acid attachment site on the tRNA molecule. This process is known as **activation** and requires ATP. The structure of a tRNA molecule and the process of activation are shown in Figure 5.

Knowledge check 5

How does the mRNA travel from the nucleus to the cytoplasm?

Examiner tip

Mixing up what occurs in transcription and translation is a common mistake. For each stage in each process make sure that you are clear on its location, the enzymes involved and the product or products formed.

Knowledge check 6

Where does translation occur?

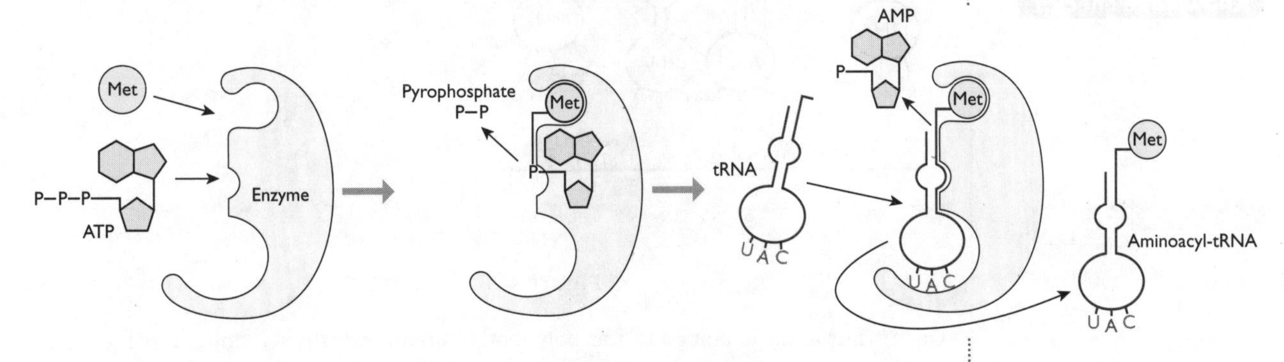

Figure 5 The enzyme shown here only accepts methionine and its specific tRNA molecule

Translation occurs at the ribosome. The two subunits of the ribosome fit around the mRNA and the mRNA now sits in the mRNA groove.

- Ribosomes have two binding sites. This means that two tRNA molecules can bind with the ribosome at any one time.
- The two amino acids carried by the two tRNA molecules become connected by a peptide bond formed by a condensation reaction and catalysed by a ribosomal enzyme.

- The first tRNA molecule then leaves the ribosome and the second tRNA moves along to take its place.
- Another tRNA molecule with the next amino acid in the sequence then fills the vacant binding site.
- The codon on the mRNA is complementary to the anticodon on the newly arrived tRNA molecule; they briefly form a codon–anticodon complex.
- The ribosome moves along the mRNA, reading each codon and the specific amino acid is added to the growing polypeptide chain. Translation stops when a stop codon is reached. At this point the two subunits of the ribosome separate and the polypeptide chain leaves the ribosome.

The process of translation is shown in Figure 6.

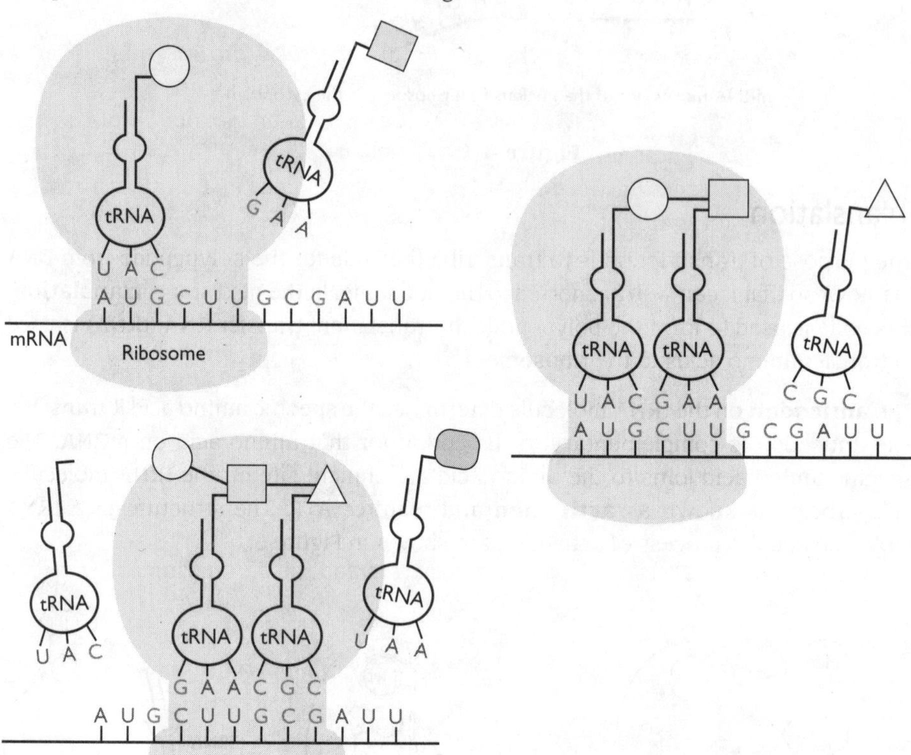

Figure 6 Translation

Once translation is complete the polypeptide chain is further modified in the Golgi apparatus. From here secretory proteins are budded off in vesicles. They then travel to the plasma membrane of the cell, the vesicle fuses with the plasma membrane and the protein is released from the cell by exocytosis.

Meiosis

Meiosis is a form of cell division that is used to produce **gametes** (sex cells) in most sexually reproducing organisms. Four haploid (containing half the number of chromosomes of an ordinary diploid body cell) daughter cells — the gametes — are produced which are genetically different from each other and from the original cell.

Knowledge check 7

Which part of a tRNA molecule is complementary to a codon on mRNA?

Examiner tip

To help you with this section you should revise the work on protein structure and the organelles involved in protein synthesis and modification from BY1.

As the gametes are haploid, when they fuse during fertilisation the zygote formed has the full diploid number of chromosomes. As the gametes are genetically different from each other and from the parent cell, meiosis also introduces genetic variation into sexual reproduction.

Meiosis has two stages, meiosis I and meiosis II. As in mitosis, each of these stages is further divided into prophase, metaphase, anaphase and telophase. The key aspects of each stage are detailed below.

Examiner tip

It is important that you can both describe what happens in meiosis and draw/recognise it on a diagram. You may also be asked to identify the stages from photos of cells undergoing meiosis. The key to answering such questions is to look carefully at what the chromosomes are doing.

Meiosis I

Prophase I

The chromatin condenses to form the chromosomes and the nuclear membrane breaks down. Each chromosome is made up of two identical **chromatids** joined by the centromere. The chromosomes are arranged in their homologous pairs. The two chromosomes in a homologous pair form a **bivalent**. The homologous chromosomes touch at points known as **chiasmata** where **crossing over** occurs. Crossing over is the swapping of genes between homologous chromosomes in a bivalent. Crossing over is an important source of variation in meiosis as it means that the two chromatids making up each chromosome are no longer identical.

The centrioles also begin to produce the protein microtubules that will form the spindle.

Metaphase I

The bivalents arrange themselves at the equator of the cell. This arrangement is random and so leads to **independent assortment**, which is another important source of variation in meiosis. The spindle fibres attach to the centromeres of each chromosome.

Anaphase I

The spindle fibres contract, the bivalents separate and the chromosomes are pulled to the opposite poles of the cell. The chromosomes are still double structures at this point as the centromere has not split.

Knowledge check 8

During which stage of meiosis do the bivalents separate?

Telophase I

Cytokinesis occurs and two new cells are produced.

Meiosis II

Meiosis II then occurs; meiosis II is similar to mitosis.

Prophase II

The spindle forms at right angles to the spindle in meiosis I.

Metaphase II

The chromosomes line up at the equator. The spindle fibres attach to the centromere.

Anaphase II

The spindle fibres contract, the centromeres split and the chromosomes are pulled to the opposite poles of the cell; they are no longer double structures.

Telophase II

The chromosomes uncoil to form chromatin and the nuclear envelope reforms. Cytokinesis occurs and each cell produces two cells.

The final product of meiosis is four daughter cells that are genetically different to each other and to the original parent cell. The process of meiosis is shown in Figure 7.

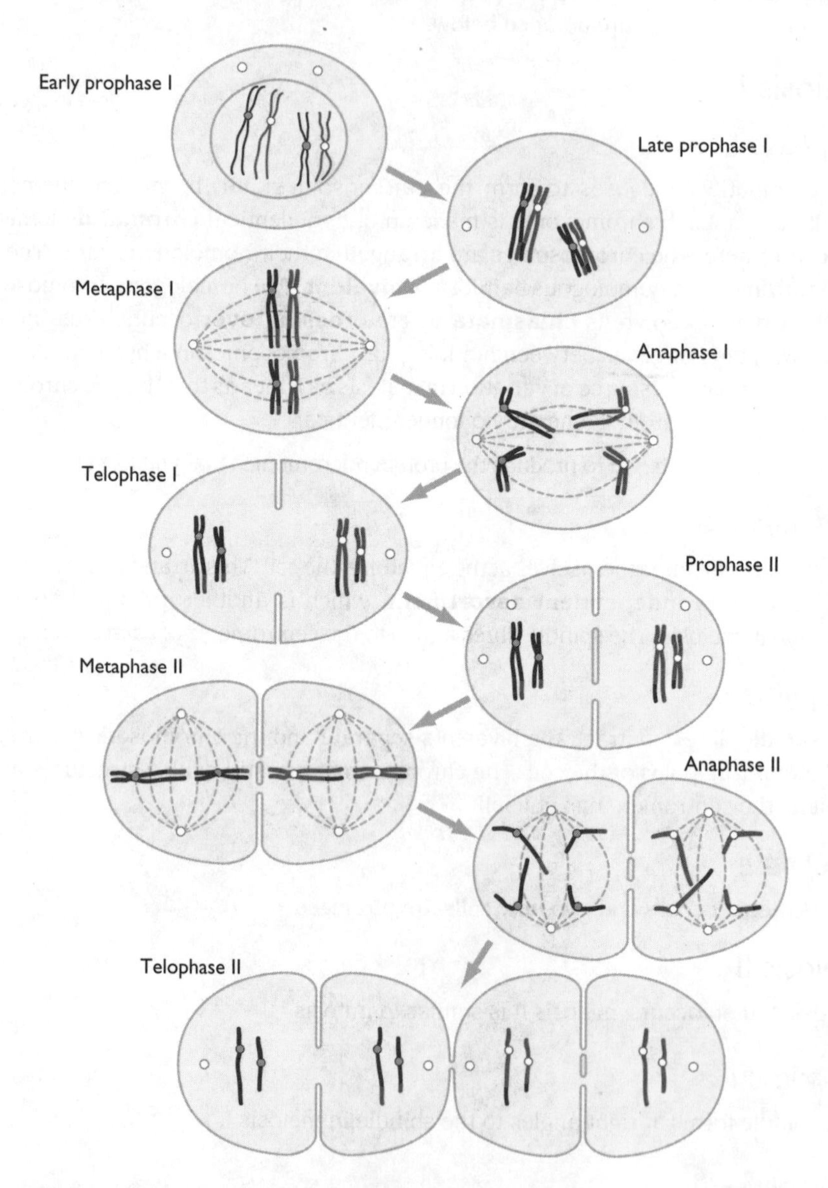

Figure 7 Meiosis to form sperm cells in an animal that has a diploid number of 4. Maternal chromosomes are blue, paternal chromosomes are red

The sources of variation in meiosis are:

- crossing over
- independent assortment
- mixing of parental genotypes

Crossing over occurs during prophase I when homologous chromosomes in a bivalent swap genes at points called chiasmata (see Figure 8).

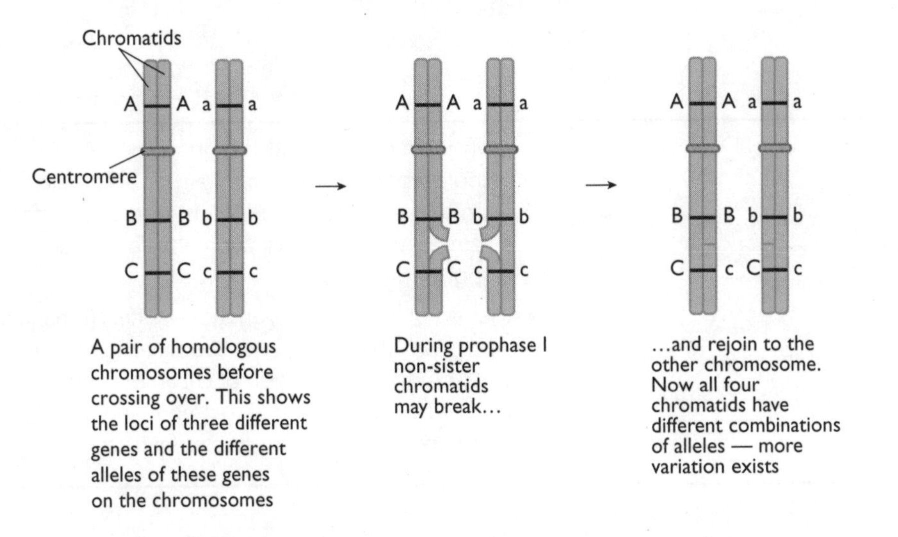

Figure 8 Crossing over produces new combinations of alleles

Knowledge check 9

Where on chromosomes does crossing over occur?

Independent assortment occurs during metaphase I when homologous chromosomes in bivalents arrange themselves randomly at the equator (see Figure 9).

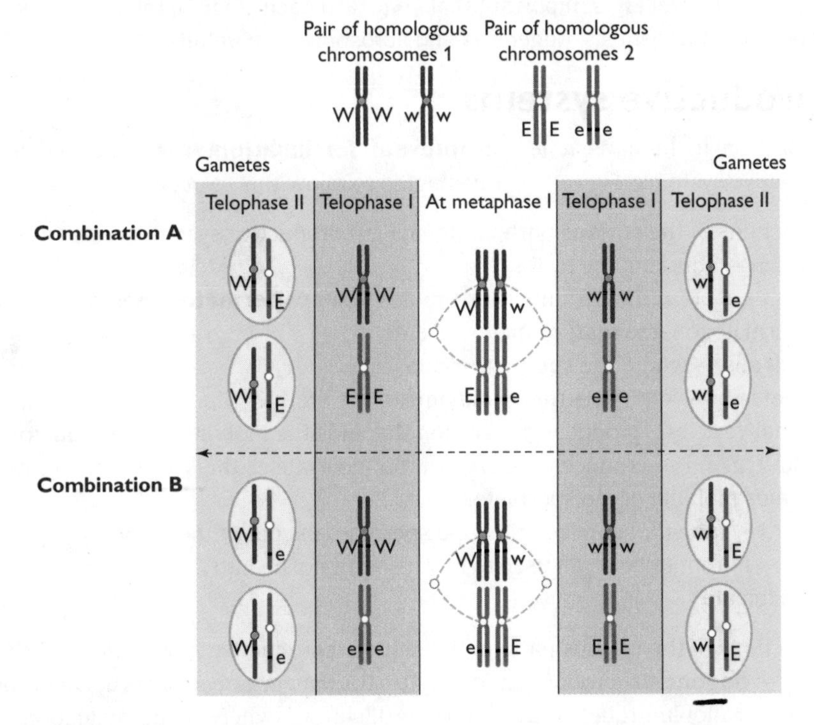

Figure 9 This diagram shows how variation can be caused by independent assortment. The different arrangement of the two pairs of chromosomes during metaphase I in A and B has led to the formation of different daughter cells

Examiner tip
Make sure you can explain fully the causes of variation in meiosis. It is a common exam question.

Mixing of parental genotypes occurs at fertilisation. When the haploid gametes fuse the chromosomes of both parents (the genotypes) are mixed to form the diploid zygote.

Summary

- In DNA replication identical copies of DNA are produced.
- DNA replication is semi-conservative. The evidence for this was provided by Meselson and Stahl's investigation using heavy (^{15}N) and normal (^{14}N) isotopes of nitrogen.
- Protein synthesis occurs in two stages:
  - Transcription occurs in the nucleus. A strand of mRNA is formed using DNA as a template.
  - Translation occurs at the ribosomes. A polypeptide is made from amino acids using mRNA as the code. tRNA is used to transfer amino acids to the ribosome to form the polypeptide chain.
- Meiosis produces four genetically different haploid gametes. Variation is produced through crossing over, random distribution of chromosomes leading to independent assortment and the mixing of parental genotypes at fertilisation.

Sexual reproduction in humans

Human reproduction is a straightforward topic with students normally answering these questions well. It is important that you learn each stage of the processes — for example spermatogenesis, oogenesis and acrosome — in detail.

Reproductive systems

Reproduction in humans relies on **internal fertilisation**. The male and female reproductive systems (Figure 10) are adapted to allow this to occur.

The functions of the various parts of the male reproductive system are as follows:

- scrotum — contains the testes
- testes — contain the seminiferous tubules where **spermatogenesis** (production of spermatozoa, the male gametes) occurs
- epididymis — where the spermatozoa mature
- vas deferens — connects the epididymis to the urethra
- seminal vesicle — produces a secretion that aids the mobility of spermatozoa
- prostate gland — produces a secretion that neutralises the alkali of the urine and also aids mobility of the spermatozoa
- urethra — tube that carries urine and spermatozoa out of the body
- penis — intromittent organ used to insert spermatozoa into the reproductive system of the female

The functions of the various parts of the female reproductive system are as follows:

- ovary — **oogenesis** (production of oocytes, the female gametes) occurs in the ovary
- oviduct (Fallopian tube) — the site of fertilisation, (where a spermatozoon fuses with an oocyte). At fertilisation a zygote is formed, which then moves down the oviduct towards the uterus.
- uterus — the embryo implants in the lining of the uterus (the endometrium) and then continues to develop in the uterus
- vagina — during sexual intercourse spermatozoa are deposited at the top of the vagina

(a)

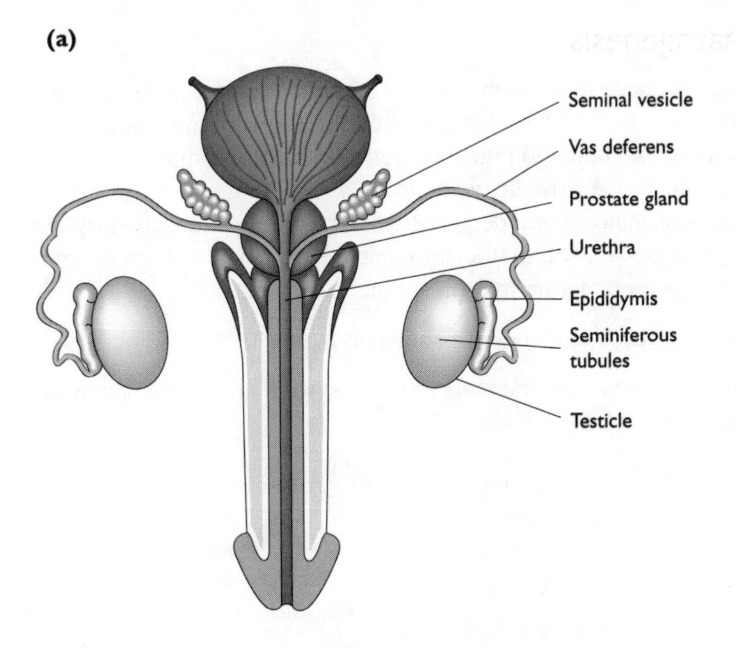

(b)

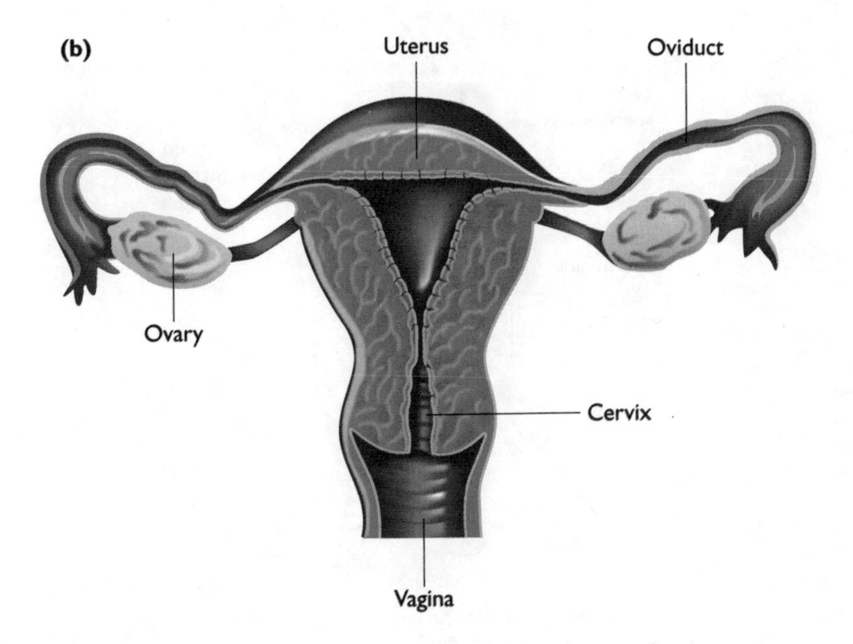

Figure 10 The structures of (a) the male and (b) the female reproductive systems

Gametogenesis

Gametes are haploid cells that contain half the full, diploid number of chromosomes. A gamete contains one chromosome from each homologous pair. The production of gametes is known as **gametogenesis**. Spermatogenesis is the production of spermatozoa (more than one sperm cell) and occurs in the seminiferous tubules in the testes; oogenesis (production of oocytes) occurs in the ovary.

Spermatogenesis

Sperm are produced in the seminiferous tubules. The process is described below:

(1) Germinal epithelial cells undergo mitosis to form spermatogonia.

(2) Spermatogonia undergo mitosis to form primary spermatocytes. Up to this point all the cells formed are diploid (contain the full number of chromosomes).

(3) Primary spermatocytes undergo meiosis I to form haploid secondary spermatocytes. Haploid secondary spermatocytes complete meiosis II to form spermatids.

(4) Spermatids then mature to form sperm.

Knowledge check 10

Is a spermatid a haploid cell or a diploid cell?

The process of spermatogenesis is shown in Figure 11.

Sperm are protected from the male immune system and nourished by Sertoli cells in the seminiferous tubules.

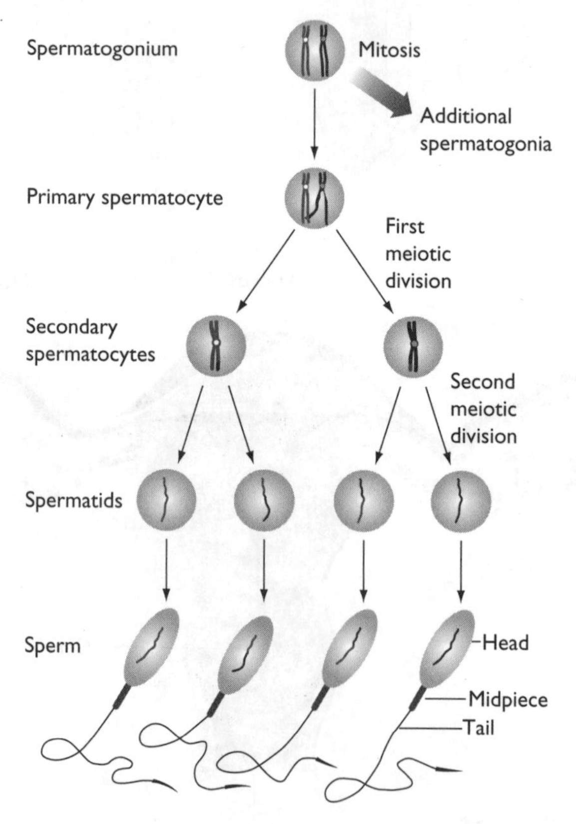

Figure 11 The process of spermatogenesis

Knowledge check 11

Where does spermatogenesis occur?

Knowledge check 12

What is the function of the Sertoli cells?

Oogenesis

Oogenesis produces oocytes in the ovaries. The first stage of the process occurs before birth:

- Oogonia divide by mitosis to form primary oocytes. The primary oocytes begin meiosis but stop at prophase I.
- Germinal epithelial cells also divide to form follicle cells which surround the primary oocyte.

The next stage of oogenesis occurs once a month once the girl reaches puberty:

- Several follicles begin to develop but only one matures into a Graafian follicle.
- The primary oocyte completes meiosis I to form a haploid secondary oocyte and a polar body.

In oogenesis, both meiosis I and meiosis II involve an uneven splitting of the cytoplasm. It is important that the oocyte contains as much cytoplasm as possible to provide nourishment for the developing embryo until it reaches the endometrium. This is achieved by polar bodies being formed in both meiosis I and meiosis II. Polar bodies are haploid so contain one-half of a full set of chromosomes, but they are very small and cannot be fertilised.

- The mature Graafian follicle moves to the surface of the ovary where it releases the secondary oocyte — this is ovulation.
- The secondary oocyte then begins meiosis II but stops at metaphase II.
- If a sperm cell enters the oocyte and fertilisation occurs, the secondary oocyte completes meiosis II, forming an ovum and a second polar body.
- The nucleus of the sperm fuses with the nucleus of the secondary oocyte to form the zygote.

Knowledge check 13

In addition to a primary oocyte what is also produced at the end of meiosis I in oogenesis?

Examiner tip

Make sure that you can state which cells in gametogenesis are haploid and which are diploid.

Fertilisation

During sexual intercourse spermatozoa are released from the epididymis. They travel up the vas deferens and are ejaculated through the urethra. Secretions from the seminal vesicle aid spermatozoa mobility and secretions from the prostate gland neutralise the alkali of any urine in the urethra. The spermatozoa are deposited at the top of the vagina. They swim up out of the vagina through the cervix along the lining of the uterus and into the oviduct where they meet the secondary oocyte which has been released from the ovary.

Knowledge check 14

Where does fertilisation occur?

Before a sperm cell can fertilise the oocyte, **capacitation** must occur. This is a biochemical process that occurs several hours after the sperm enter the female reproductive tract. The membrane surrounding the **acrosome** in the head of the sperm destabilises and prepares for the acrosome reaction that will occur when the sperm attempt to enter the oocyte.

Knowledge check 15

What must occur before a sperm cell can fertilise an oocyte?

Once capacitation has occurred the sperm cell is ready to enter the oocyte. When a sperm comes into contact with the outer jelly layer of the oocyte the acrosome reaction is triggered. The acrosome membrane ruptures and enzymes are released. These enzymes digest their way through the corona radiata and the zona pellucida, which are the outer layers around the oocyte. The head of the sperm then detaches from the tail and enters the oocyte. Only one sperm enters the oocyte. To prevent any further sperm entering, the structure of the zona pellucida changes to form a fertilisation membrane. The entry of the sperm head stimulates the secondary oocyte to complete meiosis II. The haploid nucleus of the sperm then fuses with the haploid nucleus of the oocyte to form the diploid nucleus of the zygote.

Knowledge check 16

What prevents more than one sperm from fertilising an oocyte?

After fertilisation has occurred the zygote moves down the oviduct, rapidly dividing by mitosis to form a hollow ball of cells known as the blastocyst. This rapid cell division is called cleavage. When the blastocyst reaches the uterus it implants in the uterus lining (the endometrium). The outer layer of the blastocyst is called the chorion and the outer layer of the chorion is called the trophoblast. The chorion develops chorionic villi that absorb nutrients through the endometrium. The long thin shape of the villi means that they have a large surface area. The chorion also releases human chorionic gonadoptrophin (hCG), which maintains the corpus luteum throughout pregnancy. The hormone hCG is also the basis of many pregnancy testing kits.

One potential cause of infertility in women is blockage of the oviducts. This prevents the oocyte and the spermatozoa from meeting so fertilisation is unable to occur.

Pregnancy testing

Pregnancy test kits detect hCG. This hormone is only excreted in the urine during pregnancy (women who are not pregnant do not produce hCG). Monoclonal antibodies specific to hCG are used in the test, which reduces the chance of false positives. An example of the process involved in such pregnancy test kits is outlined below:

- Urine is placed on the sample strip.
- Antibodies specific to hCG bind to hCG present in the urine. These antibodies are also attached to coloured beads. The antibodies attached to hCG are carried along the strip by the urine through capillarity.
- The antibodies attached to the hCG reach the test zone. At the test zone there are immobilised antibodies that are also specific to hCG. The hCG binds to these immobilised antibodies. Since the hCG is still attached to the first antibody, this leads to a build up of the coloured beads. This indicates a positive result.
- If there is no hCG in the urine, then the test gives a negative result.

Knowledge check 17

Which hormone is detected by pregnancy test kits?

Summary

- The human male reproductive system is adapted to introduce spermatozoa into the reproductive system of a female. The human female reproductive system is adapted to allow fertilisation to occur internally and to facilitate internal development of the embryo.
- Gametogenesis is the formation of haploid gametes. Spermatogenesis produces sperm in males and oogenesis produces oocytes in females.
- Sexual intercourse introduces sperm into the female's body. Before fertilisation can occur capacitation of sperm must occur. The acrosome reaction then allows one sperm cell to fertilise the oocyte in the oviduct, forming a zygote.
- The zygote divides to form a blastocyst, which implants in the endometrium. The chorion (outer layer of the blastocyst) develops to form the chorionic villi.
- Fertility problems can prevent couples conceiving naturally.
- Pregnancy test kits are used to detect pregnancy. They rely on monoclonal antibodies that are specific to the hormone hCG which is only produced by women during pregnancy.

Sexual reproduction in plants

There are many different types of reproduction in the plant kingdom. In BY5 you learn about sexual reproduction in flowering plants.

Pollination

The male gamete in flowering plants is contained in pollen. It is produced in the anther by meiosis. The female gamete is the egg cell produced by meiosis in the ovule. In order for internal fertilisation to occur the male gamete must come into contact with the female gamete. This occurs through **pollination**, which is the transfer of pollen from a flower of one species to the stigma of a flower of the same species.

You have to study in detail the structures of two different types of flower — those pollinated by insects and those pollinated by the wind. The structures of a wind-pollinated flower and an insect-pollinated flower are shown in Figure 12.

Knowledge check 18

Where are the male gametes produced in flowering plants?

Knowledge check 19

Define pollination.

Examiner tip

The definition of pollination is important, so make sure that you learn it.

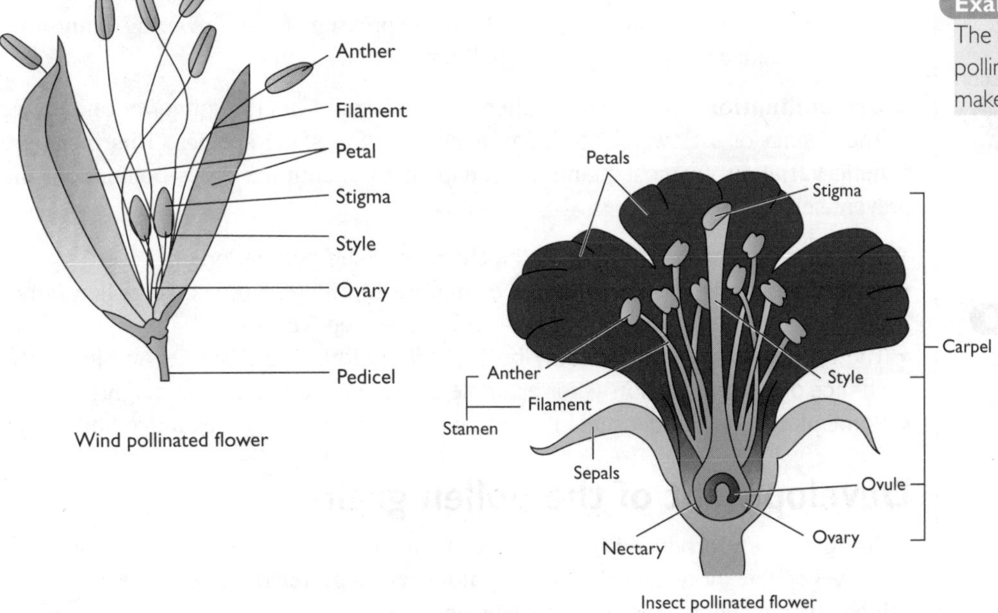

Figure 12 The structures of an insect-pollinated flower and a wind-pollinated flower

The flowers of a plant are adapted for a particular method of pollination. Insect-pollinated flowers are adapted to attract insects, to stick pollen to them from their anthers and to receive pollen on their stigmas from insects. Wind-pollinated flowers are adapted so that their pollen is dispersed by the wind and also to catch pollen that is being blown by the wind.

Insect-pollinated flowers:

- Brightly coloured petals, nectar and scent — attract insects to the flower
- Small quantities of sticky pollen — when an insect enters the flower the sticky pollen attaches to the body of the insect

- Anthers and stigma are inside the flower — this increases the chance of the insect picking up the pollen from the anther when it enters the flower and also increases the chance of pollen from the flower of a different plant being deposited on the stigma

Wind-pollinated flowers:

- Small flowers with no scent and dull coloured petals — insect attraction not needed
- Anthers hang outside the flower — increases the chance of wind blowing the pollen from the anther
- Large feathery stigma — large surface area for catching pollen blown by the wind
- Large volumes of lightweight pollen produced — the large volume increases the chance of pollination occurring and the light weight increases the chance of pollen being carried by the wind

Examiner tip

A comparison of the features of insect-pollinated and wind-pollinated plants is commonly asked for, so make sure that you understand this and can explain the differences.

There are two types of pollination — self-pollination and cross-pollination.

Self-pollination occurs when the pollen from a flower lands on the stigma of the same flower or a different flower of the same plant. Self-pollination reduces genetic variation as both gametes are from the same individual. This increases the chance of negative recessive characteristics being expressed. Most flowering plants are adapted to reduce the amount of self-pollination that occurs.

Cross-pollination occurs when pollen is transferred from the anther of one flower to the stigma of a flower of a different plant of the same species. This increases genetic variation so most plants are adapted to encourage cross-pollination and prevent self-pollination.

Examiner tip

Think about how cross-pollination and self-pollination link into the topics of variation and evolution. Cross-pollination is more likely to increase variation and therefore encourage natural selection and evolution.

There are several ways of reducing the chances of self-pollination:

- Male and female parts of the flower mature at different times, so pollen is not produced at a time when the stigma would be receptive to it.
- The anther and the stigma may be arranged within the flower to minimise the chance of self-pollination — for example they could be at different heights.
- Some plants have either male flowers or female flowers — for example holly.

Knowledge check 20

Which type of pollination leads to increased variation?

Development of the pollen grain

Pollen grains are produced in the anther by meiosis. Most pollen grains contain a single vegetative (non-reproductive cell) along with a **generative cell**. The generative cell has two nuclei: a generative nucleus and a tube nucleus. The generative nucleus divides to form the male gametes while the tube nucleus forms the pollen tube (see p. 23). The wall of the pollen grain is made up of two layers, the inner intine, which is composed of cellulose, and the outer exine, which is a tough layer that prevents desiccation (drying out). This increases the chance of a pollen grain surviving the movement from the anther to the stigma of another flower.

Once the pollen grains are mature and ready to be released from the anther the anther dries out and splits open along lines of weakness. This is called **dehiscence**. The pollen grains are now exposed to the environment and can be picked up by, for example, insects.

The ovary of the female plant produces the ovule (see Figure 13). Contained within the ovule is the egg cell nucleus, which will be fertilised by the male nucleus.

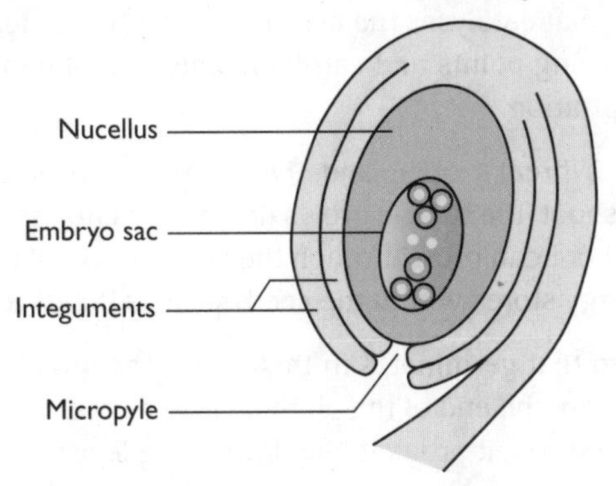

Figure 13 Cross section through an ovule

Examiner tip

For the BY5 exam you do not need to know the details of double fertilisation in plants. You may see references to this in textbooks but don't worry, you won't be asked questions on it.

The processes of pollination and fertilisation are detailed below:

- Pollination occurs when the pollen grain from a flower of one species lands on the stigma of a flower of the same species. The pollen grain then takes in water and germinates. The tube nucleus begins to form the pollen tube.
- Enzymes in the pollen tube digest their way through the style. Within the pollen tube is the generative nucleus from the pollen grain.
- The pollen tube enters the embryo sac through the micropyle (a small hole at the base of the ovule). The generative nucleus enters the ovule and fertilisation occurs — the haploid generative (male) nucleus fuses with the haploid female nucleus to form a diploid zygote.
- The seed and fruit then develop. The zygote becomes the embryo, (containing a plumule and a radicle.) The integuments form the testa or seed coat. The whole ovule forms the seed while the surrounding ovary forms the fruit, with the ovary wall becoming the wall of the fruit.

Knowledge check 21

What part of a flower becomes a seed?

Seeds are either monocotyledonous or dicotyledonous.

- Monocotyledons have only one cotyledon (seed leaf). An example is maize.
- Dicotyledons have two cotyledons. An example is broad bean.

The seeds are then dispersed. If seeds land in an area where conditions are favourable they will germinate.

Germination

Some of the requirements for germination are as follows:

- Suitable temperature — the temperature has to be close to the optimum temperature of the enzymes involved in germination. This varies depending on the species of plant.
- Water — seeds require water to mobilise their enzymes, to form vacuoles in their cells and also for transport.
- Oxygen — needed for the production of ATP by aerobic respiration.

Once the conditions for germination have been met the seed takes in water.

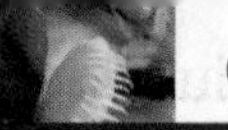

This mobilises the enzymes in the seed and allows them to break down energy-storage molecules to provide the energy for the plant to develop. An example of this is the enzyme amylase that catalyses the hydrolysis of starch to form maltose. Maltose is transported to growing points and can be further hydrolysed to produce glucose, which is used in respiration.

The seed coat (testa) breaks open and the plumule grows upwards (away from gravity) to form the shoot; the radicle grows downwards (towards gravity) and forms the roots. Until the shoot can grow through the soil surface all the energy for growth must come from energy stores within the seed, such as the endosperm.

Knowledge check 22

In a germinating seed what does the plumule develop into?

An example of a seed that germinates in this way is the broad bean, *Vicia faba*. In a germinating broad bean, the end of the plumule is folded into a 'hook' shape so that, as it grows through the soil, it protects the developing leaves. Once the plumule has grown clear of the soil it straightens, the leaves open out and photosynthesis begins.

Summary

- Plants can reproduce both sexually and asexually.
- Flowers can be pollinated by the wind or by insects. Flowers are specially adapted to their particular method of pollination.
- Cross-pollination increases genetic diversity; self-pollination reduces genetic diversity.
- Pollination is the transfer of the pollen grain from the anther of a flower to the stigma of another flower of the same species.
- The pollen grain germinates and a pollen tube grows down the style. The pollen tube enters the ovule through the micropyle and fertilisation occurs.
- The ovule forms the seed and the ovary forms the fruit.
- Plants require the correct environmental conditions for germination of seeds to occur.

Inheritance

Genetics is an important topic in BY5. On the exam, there will be a genetic cross to complete and such questions are usually worth a large number of marks. It is therefore important that you are confident in completing all the types of genetic cross that are detailed in this section.

Genetics is the study of the inheritance of characteristics. The following key terms are important:

- **Alleles** — alternative forms of the same gene
- **Homozygous** — both chromosomes in a homologous pair have the same allele. Homozygous dominant is where both chromosomes have the dominant allele; homozygous recessive is where both chromosomes have the recessive allele.
- **Heterozygous** — the alleles on the chromosomes of the homologous pair are different. The individual has a dominant and a recessive allele for a particular characteristic and could be said to be 'carrying' the recessive characteristic. This is because the characteristic is not shown by that individual but could be potentially passed on to their offspring.
- **Dominant allele** — if this allele is present, the individual has the characteristic.

- **Recessive allele** — the individual only has the characteristic if both chromosomes in the homologous pair have the recessive allele, i.e. if the individual is homozygous recessive. Cystic fibrosis is an example of a genetic condition that is caused by a recessive allele.
- **Genotype** — the alleles possessed by an organism
- **Phenotype** — the physical characteristics of an organism, controlled by the genotype

Examiner tip

These terms are key to understanding inheritance and answering questions. Make sure that you know them all.

Knowledge check 23

What is the phenotype?

The first genetic investigations were carried out by Gregor Mendel. He used pea plants with the distinct height characteristic of being either tall or short. This is an example of **monohybrid inheritance**, which is the inheritance of a pair of contrasting characteristics controlled by two alleles at a single **locus** (point on a chromosome).

In pea plants, the allele for tall, T, is dominant to the allele for short, t. A heterozygous pea plant (Tt) is crossed with a homozygous short pea plant (tt).

Parental genotypes: Tt tt

Gametes: (T) (t) (t) (t)

Offspring:

	(t)	(t)
(T)	Tt	Tt
(t)	tt	tt

Offspring genotype — Tt and tt (50% heterozygous, 50% homozygous)

Offspring phenotype — 50% tall, 50% short

Examiner tip

You must read carefully all the instructions in questions on genetic crosses. It is easy to make a simple mistake that could cost you a lot of marks.

Letters are used to represent each characteristic. The dominant allele is represented by a capital letter; the same letter in lower case is used to represent the recessive allele.

In this example, the letters to use for the different alleles are given. This may not always be the case. If you have to choose letters, pick logical ones that will not be confusing when you do the cross. It is usual to use the first letter of one of the characteristics. It is also a good idea to use letters with very different capital and lower case forms — for example G and g.

Going through the cross step by step:

(1) Show the genotypes of the parents. Remember that the parental cells are diploid, so each parental characteristic has two alleles.

(2) Represent the possible gametes from each parent. Remember that gametes are haploid so have only one allele for each characteristic (one allele from each pair). It is a good idea to draw a circle round each gamete.

(3) Construct the grid (a Punnett square) as shown above. Put the possible gametes from one parent in the first vertical column and the possible gametes from the other parent in the first horizontal row.

(4) Carry out the genetic cross. Combine the gametes and write the genotype formed in the appropriate square.

Examiner tip
It is important not to move on to more complex crosses until you are comfortable carrying out monohybrid crosses.

(5) Count the different genotypes and phenotypes and represent them as a percentage or a ratio, depending on the question. Be careful not to make a mistake when determining which genotype produces which phenotype.

Monohybrid crosses are the easiest type of genetic cross, so it is unlikely that you will be asked to carry out one in the BY5 exam. It is much more likely that you will be asked to complete one of the following types of cross, which are more challenging.

Test cross/back cross

Test crosses are used to work out whether an individual with a dominant characteristic is homozygous dominant or heterozygous. A test cross involves breeding the individual of unknown genotype with an individual that is homozygous recessive. By looking at the phenotypes of the offspring it is then possible to determine whether the unknown individual was homozygous dominant or heterozygous.

Tall and short pea plants can again be used as an example — a tall pea plant of unknown genotype is bred with a short pea plant.

If the plant were homozygous dominant, then the cross would be:

Parental genotypes: Tt tt

Gametes: (T) (t) (t) (t)

Offspring:

	(t)	(t)
(T)	Tt	Tt
(T)	Tt	Tt

Offspring genotype — Tt, 100% heterozygous

Offspring phenotype — 100% tall

If the unknown individual were heterozygous the cross would be:

Parental genotypes: Tt tt

Gametes: (T) (t) (t) (t)

Offspring:

	(t)	(t)
(T)	Tt	Tt
(t)	tt	tt

Offspring genotypes — Tt and tt (50% heterozygous and 50% homozygous recessive)

Offspring phenotypes — 50% tall, 50% short

Therefore after carrying out a test cross we can say that if breeding the individual of unknown genotype with a homozygous recessive individual produces any short plants, then the unknown individual must have been heterozygous.

Dihybrid crosses

Mendel also studied **dihybrid inheritance** in the seeds of pea plants. Dihybrid inheritance involves the inheritance of two genes not carried on the same chromosome. Each gene has two alleles. Dihybrid inheritance involves exactly the same principles as monohybrid inheritance but the presence of two sets of alleles makes it more complex. An example of Mendel's work on inheritance in pea seeds is shown below.

The seeds of pea plants can be round or wrinkled. They can also be yellow or green. Round (R) is dominant to wrinkled (r); yellow (Y) is dominant to green (y).

Therefore a homozygous dominant plant with round yellow seeds has the genotype RRYY and a homozygous recessive plant with wrinkled green seeds has the genotype rryy. A genetic cross between these two types of plant is shown below.

Parental genotypes: RRYY rryy

Gametes: (RY) (ry)

Offspring:

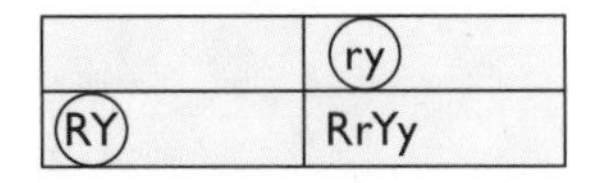

	(ry)
(RY)	RrYy

Offspring genotype — RrYy (100% heterozygous)

Offspring phenotype — 100% round, yellow seeds

These offspring are the F_1 generation (the first generation). Questions may ask you to work out the phenotypes of the F_2 generation (the second generation). These are found by crossing the offspring of the F_1 generation (two heterozygotes). For the gametes in such crosses you have to write down every possible combination of alleles. In a dihybrid cross in which the organism is heterozygous for both characteristics there are four possible gametes.

In this example, crossing plants from the F1 generation:

Parental genotypes: RrYy RrYy

Gametes: (RY) (Ry) (rY) (ry) (RY) (Ry) (rY) (ry)

Offspring:

	(RY)	(Ry)	(rY)	(ry)
(RY)	RRYY	RRYy	RrYY	RrYy
(Ry)	RRYy	RRyy	RrYy	Rryy
(rY)	RrYY	RrYy	rrYY	rrYy
(ry)	RrYy	Rryy	rrYy	rryy

This cross looks confusing at first. Just take time to think about what each genotype represents. It is important to bear in mind that when two heterozygotes are crossed the offspring phenotypes are always in the ratio 9:3:3:1 where:

- 9 — both dominant characteristics
- 3 — one dominant, one recessive characteristic
- 3 — one dominant, one recessive characteristic
- 1 — both recessive characteristics

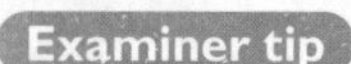

Examiner tip

Go over this method thoroughly and make sure you can do a dihybrid cross. In terms of the number of different offspring produced, a dihybrid cross is the most complex you will be asked to carry out.

Knowledge check 24

What is the expected ratio of phenotypes of offspring from a dihybrid cross between two heterozygous parents?

In this example the ratio represents:

- 9 — round, yellow seeds (RRYY, RrYY, RrYy, RRYy)
- 3 —round green seeds (Rryy, RRyy)
- 3 — wrinkled yellow seeds (rrYY, rrYy)
- 1 — wrinkled green seeds (rryy)

Mendel's laws

From his research on pea plants Mendel developed two laws:

Mendel's first law (the law of segregation) states that every organism possesses a pair of alleles for any particular characteristic. Each parent passes a copy of only one of these alleles to their offspring. Whichever of the alleles is dominant affects the phenotype of the organism.

Mendel's second law (the law of independent assortment) states that genes are passed independently of other genes from parents to offspring (the selection of a gene is not affected by the selection of other genes.) The genes assort independently during gamete formation.

Codominance

Some alleles are codominant. In this case, a heterozygous individual has a phenotype midway between the two homozygous phenotypes. An example of codominance is flower colour in snapdragons. Snapdragons that have red flowers have two alleles for red (RR); white-flowered snapdragons have two alleles for white (WW). Heterozygous snapdragons (RW) have pink flowers! This means that if a white-flowered snapdragon is crossed with a red-flowered snapdragon all the offspring will have pink flowers:

Parental genotypes: RR WW

Gametes: (R) (R) (W) (W)

Offspring:

	(R)	(R)
(W)	RW	RW
(W)	RW	RW

Offspring genotype: 100% RW (heterozygous)

Offspring phenotype: 100% pink flowers

Sex-linked inheritance

Gender is controlled by the X and Y sex chromosomes. In humans, females have two X chromosomes (XX) and males have one X chromosome and one Y chromosome (XY). A sex-linked condition is one that is controlled by an allele on the X or Y chromosome. Due to the small size of the Y chromosome most sex-linked conditions are carried on the X chromosome.

Haemophilia is an example of a genetic condition that is sex linked. It is caused by a recessive allele found on the X chromosome, represented by X^h. The implication here is that, as the allele is on the X chromosome, a female haemophiliac must be homozygous recessive (X^hX^h) while a male haemophiliac has only one recessive allele (X^hY). This is because the Y chromosome does not have this allele, so the recessive X^h allele is expressed in the phenotype. This is the reason that the vast majority of haemophiliacs are men. There are also implications for the inheritance of the condition. It is possible for a female carrier and a healthy male to have a haemophiliac son but not a haemophiliac daughter — the X chromosome from the father would be X^H as the father is not a haemophiliac. The mother of a female haemophiliac would have to be either a carrier or have the condition; her father would have to be a haemophiliac. In the cross shown below the father does not have haemophilia and the mother is a carrier (heterozygous).

Parental genotypes: X^HY X^HX^h

Gametes: (X^H) (Y) (X^H) (X^h)

Offspring:

	(X^H)	(Y)
(X^H)	X^HX^H	X^HY
(X^h)	X^HX^h	X^hY

Offspring genotypes: 25% X^HX^H (female homozygous dominant), 25% X^HX^h (female heterozygous), 25% X^HY (male dominant), 25% X^hY (male recessive)

Offspring phenotypes: 25% female non-carrier, 25% female carrier, 25% male healthy, 25% male with haemophilia

So a carrier mother and a healthy father have a 25% chance of having a son with haemophilia, but it is impossible for them to have a daughter with haemophilia.

Examiner tip

Make sure that you understand the implications of recessive sex-linked conditions being carried on the X chromosome and can answer questions on the chances of a particular pair of parents having a child of a particular gender with the condition.

Chi-squared test

Genetic crosses give the probabilities of the genotypes and phenotypes of offspring. It is important to realise that these are just probabilities and *not* certainties. For example, the chance of having a boy or a girl is always 50%, but there are many parents who have a number of daughters and no sons, or sons and no daughters. It is important to be able to find out whether the observed results from a genetic cross are significantly different from the expected results of the cross. If there is a significant difference it suggests that either the genetic cross was wrong or there is some unknown factor affecting the inheritance of the alleles. A statistical test such as the chi-squared test can be used for this purpose.

The first step is to make a **null hypothesis** and an **alternative hypothesis**. When using the chi-squared test to determine the significance of the outcome of a genetic cross, the null hypothesis is that there is no significant difference between the actual and expected results of the genetic cross. The alternative hypothesis is that there is a significant difference between the actual and expected results of the genetic cross.

Now you have to apply the chi-squared equation to your results in order to calculate the chi-squared value. The equation is:

$$\chi^2 = \Sigma \frac{(O - E)^2}{E}$$

where O stands for observed (the number of offspring of the different genotypes/phenotypes) and E stands for expected (the number of genotypes/phenotype calculated based on the probability for the genetic cross). This value is then compared to the values in a chi-squared table to determine if it shows significance or not.

When calculating the chi-squared value record the data in a table:

O	E	O–E	$(O–E)^2$	$(O–E)^2/E$

Once you have calculated the chi-squared value you have to determine the degrees of freedom. This is done by working out the number of categories and subtracting 1:

$$\text{degrees of freedom} = n - 1$$

where n is the number of categories

The calculated chi-squared value is now compared with the value in a probability table. In biology, the 5% (0.05) significance level is normally used. If the calculated chi-squared value is less than the 5% significance value in the table the null hypothesis is accepted and the alternative hypothesis is rejected, i.e. there is no significant difference between the actual and expected results. However, if the calculated value is higher than the 5% significance value the null hypothesis is rejected and the alternative hypothesis is accepted, i.e. there is a significant difference between the actual and the expected results of the cross.

Here is an example of a chi-squared calculation.

In a genetics experiment, two flowers were bred together. The expected phenotype ratio was 3:1 (75% white, 25% red). Out of 100 offspring, 40 plants had red flowers and 60 had white flowers.

- Null hypothesis — there is no significant difference between the expected and the observed results
- Alternative hypothesis — there is a significant difference between the expected and the observed results

The expected number of offspring of each colour is obtained by multiplying the total number of offspring by the probability of obtaining each colour:

$$\text{expected white flowers} = 100 \times 0.75 = 75$$

$$\text{expected red flowers} = 100 \times 0.25 = 25$$

	O	E	O–E	$(O–E)^2$	$(O–E)^2/E$
White flower	60	75	–15	225	3
Red flower	40	25	15	225	9

Using the chi-squared equation:

$$\chi^2 = \Sigma \frac{(O - E)^2}{E}$$

$$\chi^2 = 3 + 9 = 12$$

The chi-squared value is 12. The degrees of freedom for these data are 2 – 1 = 1 (number of categories (red and white) minus 1). The chi-squared value is compared with the value at the 5% (0.05) probability level in Table 2.

Table 2

	Probability												
Degrees of freedom	0.99	0.98	0.95	0.90	0.80	0.70	0.50	0.30	0.20	0.10	0.05	0.02	0.01	0.001
1	0.00016	0.00063	0.0039	0.016	0.064	0.15	0.46	1.07	1.64	2.71	3.84	5.41	6.64	10.83
2	0.02	0.04	0.10	0.21	0.45	0.71	1.39	2.41	3.22	4.60	5.99	7.82	9.21	13.82
3	0.12	0.18	0.35	0.58	1.00	1.42	2.37	3.66	4.64	6.25	7.82	9.84	11.34	16.27
4	0.30	0.43	0.71	1.06	1.65	2.20	3.36	4.88	5.99	7.78	9.49	11.67	13.28	18.46

The critical value at the 0.05 probability level is 3.84. The chi-squared value of 12 is higher than this so the null hypothesis is rejected and the alternative hypothesis is accepted, i.e. there is a significant difference between the observed and expected results.

Examiner tip
Don't be intimidated by the chi-squared test. Just follow the steps logically — it could be a source of several marks in an exam.

Linkage

All the genetic crosses covered in BY5 assume that the alleles are on different chromosomes and so will not be inherited together. However, alleles that are on the same chromosome will be inherited together because when the chromosome passes into a gamete the alleles are all present. Therefore, if that gamete is fertilised all those alleles will be incorporated into the genome of the offspring. This is known as **linkage**. The situation is further complicated with crossing over swapping genes between chromosomes, so the linkage can be changed.

Knowledge check 25
What process could occur during prophase I of meiosis to change the linkage of genes on a chromosome?

Mutations

A **mutation** is a change in the volume, arrangement or structure in the DNA of an organism. Mutations can either affect a particular gene or a whole chromosome.

Gene mutations

The different types of gene mutation are listed below along with an example of their effect on the following three DNA triplets:

Knowledge check 26
What is a mutation?

GAG/TAA/GTC

- **Addition** — in an addition mutation nucleotides are inserted into the DNA sequence:
  GAG/TAC/AGT
- **Deletion** — nucleotides are removed from a DNA sequence:
  GGT/AAG/TCA
- **Substitution** — a section of nucleotides is swapped for other nucleotides:
  GAG/CCC/GTC
- **Inversion** — a section of nucleotides is reversed:
  GAG/AAT/GTC
- **Duplication** — a section of nucleotides is duplicated:
  GAG/TAA/TAA

Since more than one triplet codes for each amino acid it is possible that a gene mutation may not affect the amino acid sequence in the primary structure of the polypeptide. For example, in DNA CTT and CTC both code for the amino acid leucine. This means a mutation that changes CTT into CTC will have no effect on the overall polypeptide.

In general, gene mutations are not beneficial. If a mutation produces a change in the DNA sequence of a gene, then during transcription this change will be incorporated into the mRNA molecule formed. This could then lead to the polypeptide formed in translation having an incorrect amino acid sequence, i.e. the primary structure of the polypeptide would be incorrect. This could lead to a change in the folding of the polypeptide and, therefore, to a change in the tertiary structure of the protein. An alteration in the tertiary structure of a protein could render it non-functional — for example an enzyme with an incorrectly formed active site. If the active site is no longer complementary to the substrate, then the enzyme will not be able to catalyse the reaction.

Mutant gene is co-dominant.

Sickle-cell anaemia is an example of a condition caused by a gene mutation. The mutation occurs in one nucleotide (thymine is replaced by adenine) in one gene. In sickle-cell anaemia the red blood cells have a rigid sickle shape. The change in shape and lack of flexibility of the red blood cells can lead to a variety of complications, including blockage of capillaries, which cuts off the blood supply to organs and results in organ damage. The allele that causes sickle cell is recessive (i.e. in a genetic cross it could be represented by a lower-case letter 's'). This means that only individuals who are homozygous for this allele (ss) have the condition. Individuals who are heterozygous (Ss) have what is known as sickle-cell trait, which does not usually cause any symptoms. However, it does mean that the individual is resistant to malaria. This encourages the sickle-cell allele to be passed on in areas where malaria is prevalent, as the heterozygous phenotype has a **selective advantage** (see page 34).

Chromosome mutations

Knowledge check 27

What condition does trisomy 21 cause?

Chromosome mutations can also occur. There are three main examples:

- **Change in structure** — the structure of a chromosome can alter because of errors during crossing over
- **Changes in number** — errors in meiosis can result in a gamete receiving an incorrect number of chromosomes. The embryo formed from this gamete will then also have an incorrect number of chromosomes. An example of this is trisomy 21. In this case, the zygote has an extra copy of chromosome 21. Humans with an extra copy of chromosome 21 have Down syndrome.
- **Changes in sets of chromosomes** — a gamete can have a whole extra set of chromosomes. This is known as polyploidy.

Examiner tip

It is important to remember that mutagens do not always cause mutations — rather they increase the frequency of mutation.

Mutation rates are normally very low. In general, organisms with short life cycles and more frequent meiosis show a greater rate of mutation. **Mutagens** increase the chance of a mutation occurring. Examples of mutagens include ionising radiations, UV light and X-rays, and certain chemicals, such as polycyclic hydrocarbons in cigarette smoke. A mutagen that increases the chance of developing cancer is a **carcinogen**. Mutations in **oncogenes** can result in cells dividing uncontrollably, leading to the development of cancer.

Knowledge check 28

What is a carcinogen?

Summary

- The phenotype of an organism is dictated by its genotype, i.e. the alleles that it possesses.
- A monohybrid cross shows the inheritance of a characteristic controlled by a pair of alleles.
- Test crosses are used to determine whether an organism is homozygous dominant or heterozygous.
- A dihybrid cross shows the inheritance of two characteristics each controlled by a pair of alleles. The ratio of offspring phenotypes in a dihybrid cross between two heterozygous parents is 9:3:3:1.
- Some alleles are codominant. Heterozygous offspring have a phenotype midway between the phenotypes of the homozygous parents.
- Alleles carried on the X and Y chromosomes are sex-linked. This has implications for the likely genders of sufferers of conditions, such as haemophilia, that are controlled by an allele carried on the X chromosome.
- The chi-squared test is used to determine if the observed results from a genetic cross are significantly different from the expected results.
- Mutations can cause changes in genes or whole chromosomes. Most mutations have negative consequences. Mutagens increase the chance of a mutation occurring. Mutations in oncogenes can lead to the development of cancer.

Variation and evolution

This topic deals with different types of variation in organisms and the driving forces behind natural selection. While working on this topic you should refer back to the work you did on evolution in BY2.

Variation in organisms can be separated into two categories: continuous and discontinuous variation.

Discontinuous variation

One pair of alleles usually control **discontinuous variation**. It is categoric. This means that the characteristics fit into one of several categories and are not affected by environmental conditions. Suitable examples of discontinuous variation are blood groups and earlobe attachment (see Figure 15(a)) in humans.

Continuous variation

Continuous variation is controlled by a number of different alleles. If the organism inherits alleles for this characteristic then it may be expressed in their phenotype. However, the environment plays a part (see Figures 14 and 15(b)). For example, an individual may inherit alleles for being tall. However if the individual then has a poor diet he or she will not reach their potential height. Continuous variation can have any value on a scale between a maximum and a minimum.

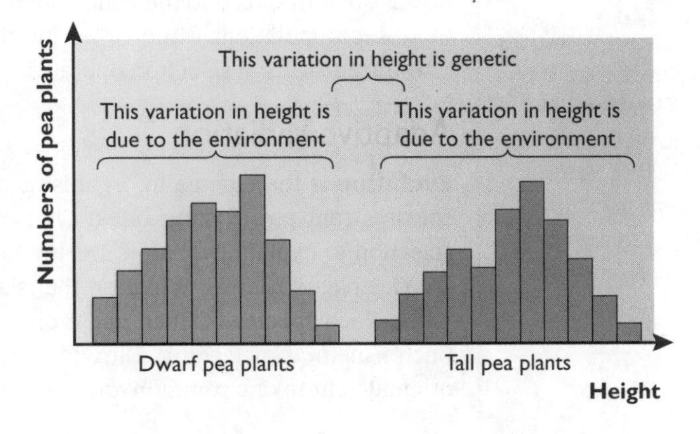

Figure 14 Genetic and environmental variation of height in pea plants

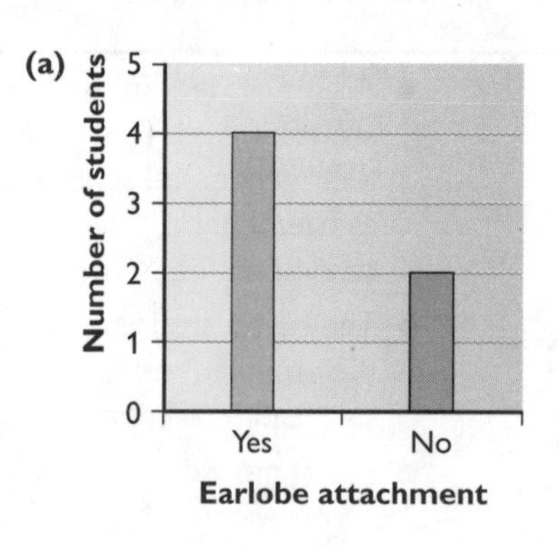

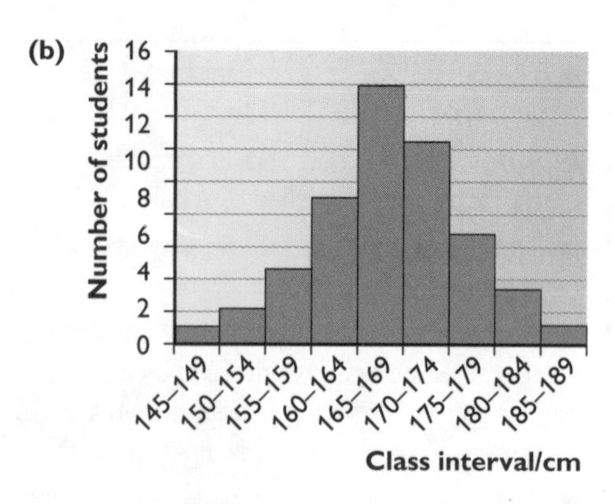

Figure 15 (a) A bar chart of earlobe attachment and (b) a histogram of height in 17-year-old students

Variation can be either heritable or non-heritable.

Environmental conditions can cause non-heritable variation. For example, differences in diet can lead to a large variation in body shape and size within a population of animals. However, this variation is not passed on to the next generation because the genes of the organism are not altered. Therefore non-heritable variation does not lead to natural selection. A variation in the genes in the gametes of an organism will be passed onto the offspring. This therefore is heritable variation.

Knowledge check 29

Not all mutations give rise to heritable variation. Explain why.

Gene pool and allele frequencies

The **gene pool** is all the alleles in any population. The environment exerts selection pressures, which regulate the frequency of an allele in the gene pool. These selection pressures can take a wide range of forms from climate to predators. The organisms that are best adapted to survive these selection pressures are said to have a selective advantage. The alleles that determine the characteristics that confer the selective advantage will be selected for in the gene pool and their frequency will increase.

Knowledge check 30

If an allele leads to a selective disadvantage in the environment what is the likely effect on the allele's frequency in the gene pool?

Gene pools are open or closed to a greater or lesser extent. In open populations, new alleles are introduced to the gene pool through interbreeding with other populations. In closed populations, little or no interbreeding occurs and new alleles are rarely introduced. Natural selection operates within gene pools.

Examiner tip

When answering questions on evolution and variation the use of key terms is important. Make sure that you include terms such as 'gene pool', 'allele frequency' 'variation' and 'natural selection' in your answers.

Adaptive radiation

Evolution is the change in organisms over many generations. It gives rise to new species from pre-existing ones. Charles Darwin developed the theory of natural selection to explain evolution. During his travels on the ship *HMS Beagle* he visited the Galapagos Islands. While on the islands he collected samples of finches (small birds). Each species of finch had a different-shaped beak adapted to that particular finch's method of feeding. Darwin hypothesised that all the species of finch had originated from one common ancestral species that had travelled over from the South

American mainland. As the Galapagos Islands were relatively newly formed there was little competition and a wide range of possible ecological niches the birds could fill. Through natural selection the birds' beak shapes and sizes changed to take best advantage of different ecological niches. This evolution of a number of species from a single common ancestor to fill a range of ecological niches is an example of **adaptive radiation**.

Knowledge check 31

What feature of Darwin's finches became adapted to their particular diets through adaptive radiation?

Natural selection

Mutations are important in **natural selection**. It is because of mutations that there is genetic variation in a population. This variation leads to some individuals being better adapted to survive in the environment than other individuals, i.e. some individuals have a selective advantage. Organisms with a selective advantage are more likely to survive and reproduce. By surviving and breeding they pass on the genes that gave them their selective advantage to their offspring. This then means that the offspring are more likely to survive. This process occurs over a long period of time and results in the characteristics that gave the selective advantage becoming widespread in the population. Through natural selection the species has evolved.

Two important aspects of natural selection are overproduction and competition. In overproduction more offspring are produced than are required to replace the parents. However, over time populations remain relatively stable. This is because of competition:

- **Interspecific competition** is competition between members of different species.
- **Intraspecific competition** is competition between members of the same species.

Knowledge check 32

Distinguish between interspecific competition and intraspecific competition.

Genetic drift

Changes in allele frequency are not always due to natural selection. Allele frequencies can also change due to chance. This is known as **genetic drift**. Genetic drift is different from natural selection in that allele frequency is changed through random chance, not because the allele confers, or does not confer, a selective advantage.

Founder effect

The **founder effect** is a specific type of genetic drift. It occurs when a small group from a population colonises a new area and forms a new population. Because of the chance make-up of the new population, there may be a reduction in genetic diversity compared with that of the original population. Over time, this reduction in genetic diversity and of different alleles within the gene pool of the new population can lead to large genotypic and phenotypic differences between the new population and the old population. The founder effect is illustrated in Figure 16.

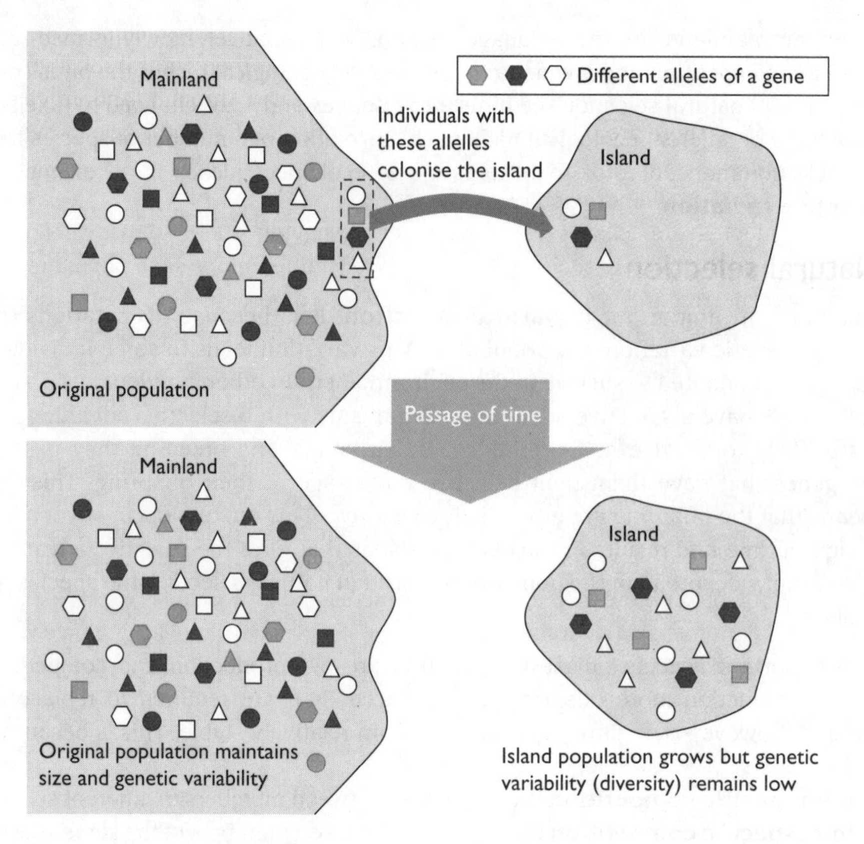

Figure 16 Consequences of the founder effect

Effect of natural disasters

A **natural disaster** can lead to extreme examples of genetic drift. If a large proportion of a population is wiped out randomly by a natural disaster then this causes a random and severe change in the allele frequency of the population's gene pool. This is an example of a genetic bottleneck, which you studied in BY2.

Speciation

Knowledge check 33

What is a species?

Speciation occurs when two groups of organisms can no longer breed together to produce fertile offspring and are, therefore, two different species.

There are two main types of speciation:

- allopatric speciation
- sympatric speciation

Examiner tip

Do not confuse allopatric and sympatric speciation — they are very different. Make sure that you learn the correct names for both types.

Allopatric speciation

Allopatric speciation occurs when populations that once interbred become separated geographically into two groups. For example, this could be due to the formation of a large river, the sealing off of a lake from a larger body of water or an earthquake. These groups are called **demes** — local populations that interbreed and share a distinct gene pool. In order for speciation to occur there can be no exchange of genes between the two demes. Over time, the gene pools of the two demes will alter. If the demes are experiencing different selection pressures then natural selection will speed up this process; however gene pools can also alter through random genetic drift. Assuming there are different selection pressures acting on the two demes the following will occur:

- There is variation in both demes.
- Selection pressure leads to some individuals having alleles that give them a selective advantage. As the selection pressure is different for each population the alleles conferring the selective advantage are also different.
- The organisms with the selective advantage survive, breed and pass on the advantageous alleles to their offspring.
- This is repeated many times over a very long period of time and through many generations.
- Eventually the genes of organisms from the two populations become so different that even if the animals were able to interbreed and produce offspring the homologous chromosomes in the offspring would be so different from each other that they would not be able to pair up during prophase I of meiosis. Therefore, this organism would not be able to produce gametes and so would be infertile. As the two populations could no longer interbreed to produce fertile offspring they would now be considered two different species.

Allopatric speciation is illustrated in Figure 17.

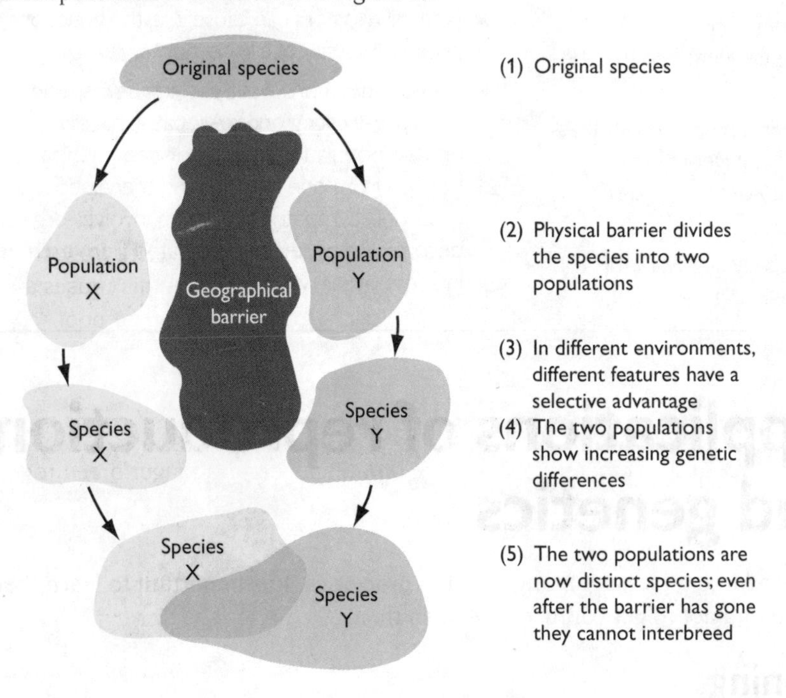

Figure 17 Allopatric speciation

Knowledge check 34

What is the term for speciation caused by a physical barrier that prevents two groups of organisms from interbreeding?

Sympatric speciation

Groups of organisms within a population can be isolated in other ways, not just geographically. Speciation that arises from these different forms of isolation are

given the general name **sympatric speciation**. Possible isolating mechanisms in sympatric speciation include:

- **Mechanical isolation** — variation in the sexual organs of organisms of the same species can lead to them being unable to mate successfully. This type of isolation is seen in both insects and plants.
- **Behavioural isolation** — variations in courtship and mating behaviour can lead to groups of a particular species becoming isolated from one another. This has been observed in the species of fruit fly *Drosophila*.
- **Gametic isolation** — although the gametes of two different organisms have the potential to meet, fertilisation does not occur. This type of isolation is common in marine invertebrates.
- **Hybrid inviability** — fertilisation occurs however the embryo is unable to develop into a living organism.
- **Hybrid sterility** — a hybrid organism is formed, but it is sterile because its homologous chromosomes are unable to pair up during meiosis I, so gametes cannot form.

Knowledge check 35

Why are hybrid organisms sterile?

Summary

- Variation can be continuous or discontinuous and heritable or non-heritable.
- Charles Darwin developed the theory of evolution through natural selection.
- A gene pool is all the alleles in a population. Allele frequency is a measure of the frequency of an allele in a population.
- A selection pressure is an environmental effect that leads to some alleles being advantageous. Through natural selection, the frequency of these alleles within the gene pool increases.
- Allele frequencies in gene pools can also vary through random chance — this is known as genetic drift.
- A small group from a population establishing a new population may be unrepresentative of the gene pool of the original population. This is known as the founder effect.
- Natural disasters can cause drastic reductions in the genetic diversity of a gene pool.
- Speciation is a process by which new species arise from pre-existing ones. Allopatric speciation relies on two populations becoming geographically isolated from one another and then speciation occurring. Sympatric speciation involves organisms becoming reproductively isolated from each other in some other way.

Applications of reproduction and genetics

This is a large topic with many different processes. It is important to learn the detail of each and not to get confused between them.

Cloning

Clones are organisms that are genetically identical to each other. Clones can be produced naturally by asexual reproduction, for example in bacteria or by plants with the formation of suckers or bulbs. Humans can produce clones artificially. Outlined below are the different methods of producing clones.

Knowledge check 36

What are clones?

Micropropagation

Plants can be cloned from just a few cells by micropropagation. This relies on the fact that many plants cells are **totipotent**, which means that they can divide to form the other cells of the plant. The key steps in micropropagation are:

- A sample of cells is removed from the meristem (growing region) of a plant and is divided to form explants.
- The explants are placed in a sterile, aerated medium.
- The cells of the explant divide by mitosis to form a callus.
- The callus is subdivided and each piece differentiates into a plantlet.
- The plantlets are then transplanted into sterile soil.

The advantages of micropropagation are that:

- the plants all share the same desirable characteristics, such as high product yield
- the survival rates of plants are high, due to the sterile, controlled conditions
- storage and transport are more efficient as large numbers of plants can be stored and transported together, which reduces overall costs

There are however disadvantages to micropropagation:

- Contamination of the culture medium by fungi or bacteria may result in the loss of large numbers of plants. This means that sterile conditions are essential.
- There is an increased risk of mutation among the cloned plants. They must be carefully checked and defective plants removed. This increases overall costs.

Animal cloning

The two main methods of animal cloning are embryo cloning and nuclear transfer.

Embryo cloning involves the splitting up of the cells that make up the embryo. As these undifferentiated cells are totipotent they can divide to form an embryo that can then be implanted into a surrogate mother. Organisms produced by embryo cloning will not be identical to the mother or father but will be genetically identical to each other. This is useful to farmers who want to quickly produce large numbers of offspring from animals with desirable characteristics.

The method is as follows:

- An embryo is produced by adding spermatozoa from an organism with desirable characteristics to an egg from a female of the same species that also has desirable characteristics.
- Once fertilisation has occurred the zygote is left to divide by mitosis to form a ball of cells.
- This ball of cells is then divided and implanted into several different surrogate mothers.
- When the offspring are born they will be genetically identical to each other and will hopefully share the beneficial characteristics of their parents.

Nuclear transfer cloning is used to produce a clone of an already living animal:

- A somatic (body cell) is taken from the animal to be cloned.
- The nucleus is removed.
- An oocyte from an animal of the same species is also taken and its nucleus removed.
- The nucleus from the animal to be cloned is fused with the oocyte using an electric shock.

- This oocyte will now develop into a zygote that will be implanted into a surrogate mother.
- The organism born will be identical to the animal from which the original nucleus was taken.

Animal cloning is illustrated in Figure 18.

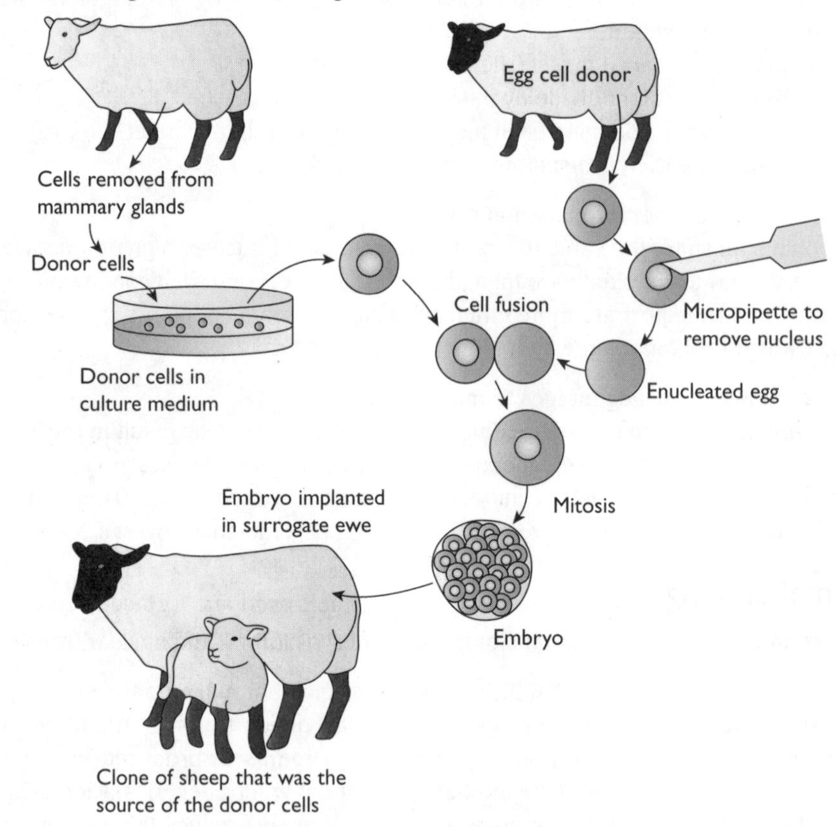

Figure 18 The steps involved in cloning sheep and cattle

Examiner tip
Make sure you know the differences between the methods of cloning. Students often mix them up in exams.

Cell culture can also be used to produce large numbers of identical cells that can then be used for research purposes. For example, a large number of cloned tumour cells could be produced and anti-cancer drugs could be tested on them.

The cloning of **stem cells** can be used in **tissue engineering**. Stem cells are undifferentiated cells. This means they have the potential to divide and differentiate into the original organism's specialised cells. These stem cells can then be used to produce tissues or organs. There is a lot of ethical debate about the use of stem cells because they are mostly derived from embryos and the embryo is destroyed in the process.

Knowledge check 37
What is a stem cell?

Advantages and disadvantages to cloning

Advantages include:

- Large numbers of organisms can be produced quickly — this is particularly true of embryo cloning in animals.

- As the clones produced are genetically identical they will share the same desirable characteristics.

Disadvantages include:
- Cloning can be expensive and unreliable, particularly the cloning of mammals.
- Ii is possible that the clones will possess disadvantageous alleles and these could lead to long-term unforeseen effects.
- As cloning of humans is considered unethical, it cannot be carried out even for possible beneficial reasons.

Examiner tip

When describing advantages and disadvantages to any part of applications of genetics, make sure you are specific. You won't get marks for terms such as 'unethical', 'playing God' or 'designer babies'.

IVF

IVF (in vitro fertilisation) is a technique that can be used to help couples who are unable to conceive naturally. The process of IVF is as follows:
- Ovulation is stimulated using hormones that lead to several follicles maturing at the same time. Therefore, several oocytes are released.
- The oocytes are collected from the woman. Ultrasound is used to guide a tube through the vagina into the oviducts and the oocytes are taken.
- Semen (containing the sperm) from the male is stored in a nutrient-rich solution.
- Each oocyte is placed in a separate container along with approximately 75 000 sperm. When the sperm count is low or the sperm struggle to move (low motility), the sperm can be injected directly into the oocyte.
- After 3 days, two of the oocytes that have been fertilised and formed zygotes are implanted into the woman's uterus. Using two embryos increases the chances of at least one of them implanting successfully and developing. In some cases more than two embryos are implanted.
- IVF results in many unused embryos. These unused embryos are normally discarded, but they could be used as a source of stem cells. There is, however, a lot of ethical debate about obtaining stem cells in this way.

Human genome project

The human genome project is an international project set up to determine the sequence of nucleotides that make up the DNA of humans and to identify and map the genes it contains. As part of the work, some genes have been sequenced (their nucleotide sequence is known) and their location on the chromosomes determined (gene mapping). The complete nucleotide sequence of the human genome is now known but work is continuing to identify and determine the functions of the genes within the genome. The human genome project has a number of potential benefits including:
- It allows for the development of new and better targeted medical treatments.
- It has increased the opportunities for screening for genetic disorders. By knowing the sequence of the allele(s) that causes a genetically determined disease, scientists can determine whether a person will develop the disease.
- Scientists can also look for incidences of mutation in certain genes that may result in genetic disorders.

Genetic counselling and genetic screening

Genetic counselling

In genetic counselling patients who are at risk of developing or transmitting a genetic disease are advised of the consequences of the condition and the risks of transmitting it to their offspring. A genetic counsellor helps diagnose genetic disorders and also supports patients who have them.

When advising people on the risk of having children who may have a genetic disorder, genetic counsellors will consider:

- the number of people with the condition in the general population
- whether the parents are closely related
- whether either parent has a family history of the condition

The genetic counsellor is then able to advise if screening is appropriate.

Genetic screening

Genetic screening has a range of potential benefits:

- It can be used to determine if an individual is a carrier (has a recessive allele) for a particular condition.
- It can be used to screen unborn children for genetic diseases. Examples of tests include:
  - **blood tests**
  - **amniocentesis** — removing a sample of amniotic fluid. This will contain some cells from the fetus, which can then be tested.
  - **chorionic villi sampling** — a very small volume of tissue is removed and the cells tested. Chorionic villus sampling is used to screen for cystic fibrosis.
- It can be used to test children or adults for conditions that have not yet developed. This could include Huntington's disease, which usually only produces symptoms in middle-age, or to determine whether an individual has an increased risk of developing some cancers.
- Genetic testing can be used in diagnosis and in forensic testing to determine identity.

There are also potential disadvantages to genetic screening:

- Genetic screening for some conditions such as cancer only gives an indication of increased risk. This means that it can be difficult to interpret results and may lead to people who will never actually develop the disease becoming anxious.
- There is a risk of the tests being used to discriminate against people. The results of tests could be used when making decisions about employing people or in financial matters, such as insurance or mortgages.
- There is a risk of false positives (people being told they have an allele for a disease when this is not the case) or non-diagnosis of a condition due to laboratory error.

Gene therapy

Knowledge check 38

What is gene therapy?

Gene therapy is the process of replacing a faulty gene that is causing a genetic disease with a healthy gene. There are two methods of gene therapy:

- **Germ-line gene therapy** — replacing the genes in the spermatozoa or the egg; this is currently banned in most countries

- **Somatic cell gene therapy** — the genes are replaced in the patient

For gene therapy to work, the healthy gene must be introduced into the cells of the patient, transcription and translation must occur and the correct non-faulty protein produced. This is a complex and difficult process. At the moment there are no gene therapy treatments in use. However, a treatment for a genetic disease that can lead to acute pancreatitis is about to be approved. One of the main problems in gene therapy is getting the gene into the patient's cells. A **vector** is used to do this. Examples of vectors include viruses and liposomes.

Knowledge check 39
How does somatic-cell gene therapy differ from germ-line gene therapy?

Cystic fibrosis

Currently trials are underway into treating cystic fibrosis with gene therapy.

Cystic fibrosis is a condition caused by a faulty cystic fibrosis transmembrane regulator protein (CFTR protein). CFTR transports chloride ions out of epithelial cells into the surrounding mucus. This leads to a lowering of the water potential of the mucus. Water then moves out of the epithelial cells into the mucus by osmosis. This ensures that the mucus remains thin and not viscous.

In a person with cystic fibrosis the DNA sequence that codes for the CFTR protein is incorrect (the most common mutation is a deletion of one amino acid). This means that the protein does not form correctly and is non-functional. Without the CFTR protein, chloride ions are not pumped out of the epithelial cells, water does not move into the mucus by osmosis and the mucus, therefore, becomes thick and sticky. This causes blockages in the respiratory, digestive and reproductive systems, which lead to infections and inflammations. These cause structural changes to the lungs and cardiorespiratory problems that eventually lead to death. Current treatments for cystic fibrosis include a range of drugs to reduce the chance of infections and alleviate other symptoms, as well as daily physiotherapy.

Knowledge check 40
What ions are transported by the CFTR protein?

Gene therapy could be used to replace the CFTR protein. The process is as follows:

- The cystic fibrosis gene from a healthy donor is identified using a **gene probe**. This is a short section of single-stranded DNA that is labelled in some way. It is complementary to the sequence of DNA in the gene being looked for and binds to it, indicating the presence of the gene.
- The gene is cut out using **restriction enzymes**.
- Many copies of the gene are made.
- The gene is then placed in a liposome (a sphere of phospholipid molecules).
- The liposome is introduced into the patient's respiratory system using an inhaler.
- The liposome fuses with the membranes of the epithelial cells and the gene travels to the nucleus.
- The new gene is incorporated into the cell's DNA. If the new gene undergoes transcription and translation the correct CFTR protein will be produced. The symptoms of cystic fibrosis should then be alleviated.
- However, continued treatment is required as the next generation of epithelial cells that are produced will still have the faulty gene.

While gene therapy has the advantage of potentially providing treatment for serious conditions such as cystic fibrosis there are some possible disadvantages. These include side effects such as adverse reactions to the vector and possible activation of oncogenes.

Examiner tip
Cystic fibrosis and its treatment by gene therapy are complex, so make sure you understand both fully.

Genetic engineering

Genetic engineering utilises **recombinant DNA** technology. Recombinant DNA is formed when a new piece of 'foreign' DNA is incorporated into a bacterial plasmid. The general process of forming recombinant DNA is as follows:

- The required gene is identified using a gene probe.
- The gene is cut out using restriction enzymes.
- These restriction enzymes produce DNA with unpaired bases — 'sticky ends' (see Figure 19). The same restriction enzymes are then used to cut a bacterial plasmid. As the same restriction enzyme is used it produces sticky ends that are complementary to the gene. The enzyme DNA ligase is then used to join (splice) the sticky ends together.

Knowledge check 41

What are sticky ends?

The transfer of a gene into a plasmid is illustrated in Figure 20.

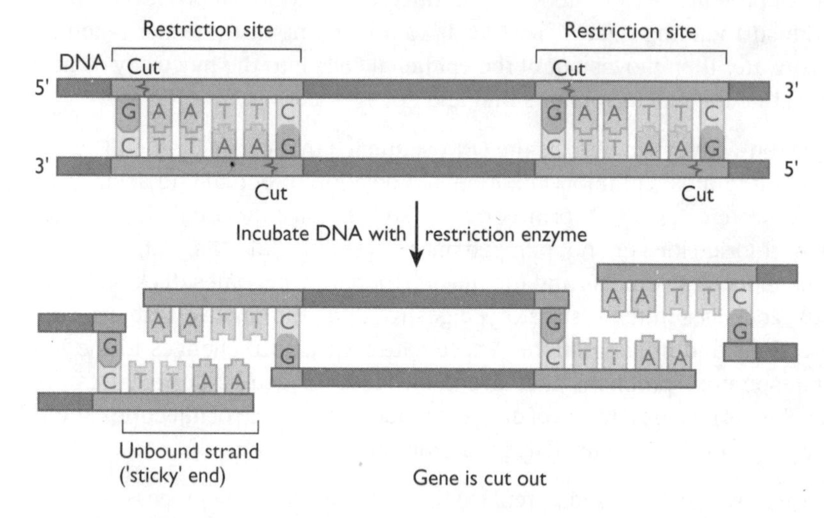

Figure 19 Staggered cuts in DNA produce 'sticky ends'

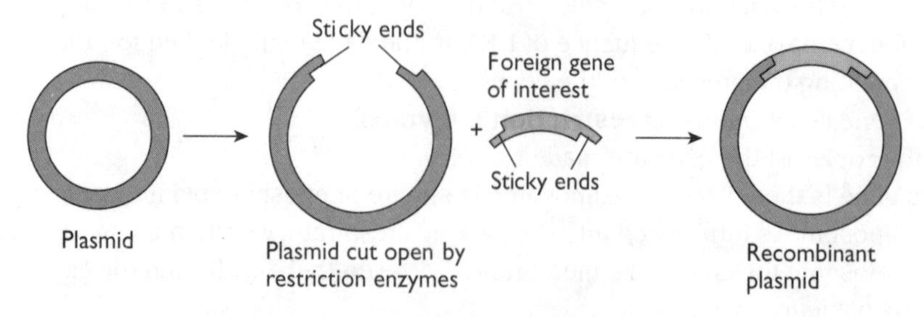

Figure 20 Transferring a gene into a plasmid

Instead of using a gene probe, reverse transcription (see Figure 21) can be used to produce a section of DNA from a strand of mRNA. The process is as follows:

- The mRNA that codes for the desired protein is extracted.
- The enzyme reverse transcriptase builds a complementary strand of single-stranded DNA (cDNA) from free DNA nucleotides.
- The enzyme DNA polymerase is then used to produce a double-stranded DNA molecule from the cDNA. This DNA will code for the required protein. The DNA can then be incorporated into a plasmid as detailed above.

Knowledge check 42

Name the enzyme used to produce a strand of cDNA from a strand of mRNA.

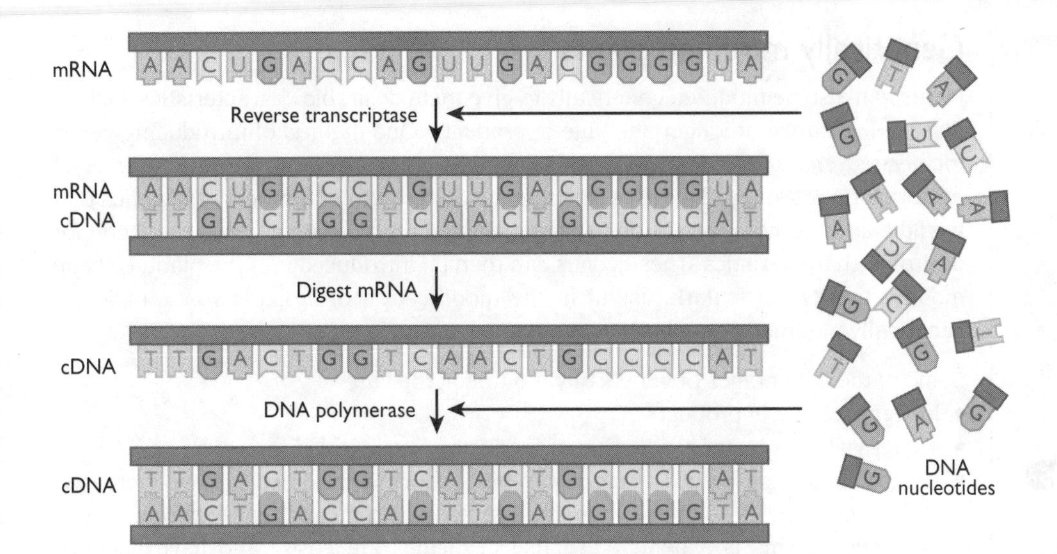

Figure 21 Creating the gene using reverse transcriptase

Genetic engineering of bacteria

An example of the use of recombinant DNA technology is the production of human insulin, used to treat diabetes, by bacteria. The process is as follows:

- The gene for insulin production is identified in a healthy human cell using a gene probe. Specific restriction enzymes are then used to cut the gene from the DNA. The same restriction enzymes are then used to cut a bacterial plasmid.
- This produces complementary sticky ends in both the insulin gene and the plasmid.
- DNA ligase is then used to join (splice) the insulin gene into the bacterial plasmid.
- A **marker gene**, such as a gene for resistance to a particular antibiotic, is also inserted into the plasmid.
- The plasmid is then introduced into a bacterial culture. Some of the bacteria will take up the plasmid. The marker gene is used to identify the bacteria that have taken up the plasmid and are producing insulin. In the case of an antibiotic-resistance marker gene, antibiotics are added. Bacteria that have not taken up the plasmid are killed.
- Surviving bacteria are cultured in a fermenter to produce a large population of bacteria, all producing insulin.
- The insulin is then extracted and purified.

Knowledge check 43

Which enzyme is used to join sticky ends of a DNA molecule?

Knowledge check 44

Why are antibiotic-resistance marker genes used in genetic engineering of bacteria?

The main advantage of using recombinant DNA is that a complex protein like human insulin can be produced in large quantities. This removes the need to use insulin from other mammals (such as pigs) in the treatment of diabetes.

There are however considerable disadvantages to using recombinant DNA to produce human products:

- It is a technically complicated process and is, therefore, expensive to carry out on an industrial scale.
- It may be difficult to identify the gene for the product, or several genes may be involved in the production of the product.
- Not all genes in eukaryotes can be expressed in prokaryotes.
- There is concern about using antibiotic-resistance marker genes that could be transferred to free-living pathogenic bacteria.

Genetically modified plants

Plants can also be modified genetically to give them desirable characteristics such as disease resistance or longer shelf life of products. One method of introducing genes for desired characteristics into plants involves using bacteria. Certain species of bacteria attack damaged plants and stimulate the growth of tumours. The genes that would lead to tumour formation are removed and replaced with genes that code for desirable characteristics. These genes can then be introduced into the plant. Genetic modification is particularly useful in the modification of crops. Two examples of genetically modified crop plants are tomatoes and soya.

Some of the advantages of genetically modified crops are:

- longer shelf life of products
- higher yield — greater production of products
- reduced use of pesticides — crops can be genetically modified to have increased resistance to pests such as fungi and insects, which means that fewer pesticides are needed. This is a benefit because pesticides are costly and have negative ecological impacts.

Knowledge check 45

Name two examples of crops that have been genetically modified.

Possible disadvantages to the use of genetically modified crops include:

- There is a risk of pollen from genetically modified plants pollinating wild natural plants. This could lead to unforeseen consequences for the natural populations.
- There are also concerns about possible long-term health impacts of eating products derived from genetically modified organisms.
- Increased use of genetically modified crops could also lead to an overall reduction in biodiversity.

Examiner tip

In a criminal investigation, an exact match of bands is required. In paternity testing half the bands have to match (half the DNA is from the father and half from the mother).

Genetic fingerprinting

Every person (with the exception of genetically identical twins) has an individual genetic fingerprint (sometimes known as a DNA profile). As genetic fingerprints are unique to an individual they can be used in forensic criminal investigations — for example to determine if a suspect was present at the scene of a crime. Genetic fingerprinting can also be used for paternity testing.

Genetic fingerprinting uses sections of non-coding DNA called **introns**, which contain blocks of repeated nucleotides. Every individual has a different number of repeating blocks. Introns are distinct from **exons**, which are the sections of DNA that code for proteins. The process of genetic fingerprinting is as follows:

Knowledge check 46

What is an exon?

- DNA is extracted from the cells in the sample.
- Restriction enzymes are used to cut the DNA into different-sized fragments. Each individual will have fragments of different lengths.
- The DNA fragments are separated using electrophoresis. The DNA is placed into small wells in a gel and a voltage is applied across the gel.
- As the DNA is negatively charged it is drawn towards the positive anode. Different length fragments of DNA move at different rates up the gel; longer fragments move more slowly than shorter fragments.
- After a given time the process is stopped. As each individual has different-sized fragments of DNA the pattern formed by the different distances moved is unique to each individual. However, the DNA is not visible.
- A technique known as Southern blotting is used to make the DNA visible.

- A nylon membrane is placed onto the gel. The DNA fragments transfer onto this membrane in their original pattern. The DNA is then fixed onto the membrane.
- DNA probes are then added to the membrane. These probes have some kind of marker. The probes may be slightly radioactive or have molecules of a dye bonded to them. The probes bind to the DNA fragments. The membrane is then washed to remove any non-bonded probes.
- If a radioactive probe is used the membrane is then placed under an X-ray film. The radioactive probes cause dark bands to appear on the film at sites where they have bonded to the DNA fragments. This produces a pattern of dark and light bands that is unique to the individual — a genetic fingerprint.

The process of obtaining a genetic fingerprint is shown in Figure 22.

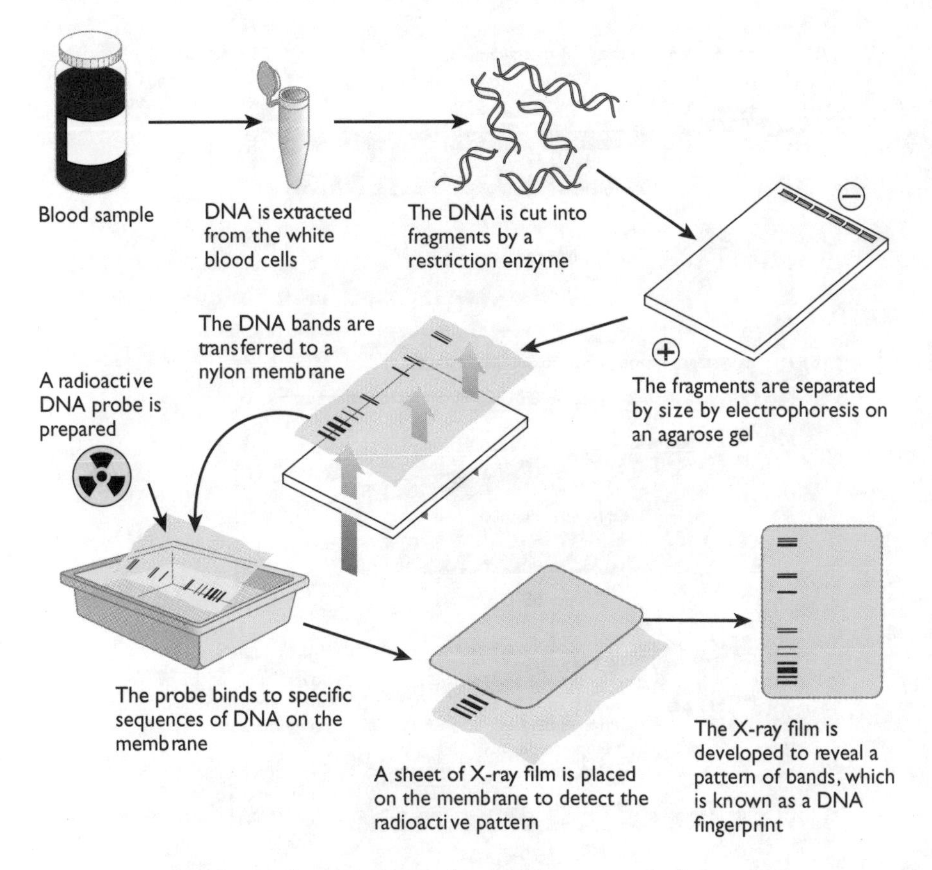

Figure 22 The main stages in obtaining a genetic fingerprint

PCR

The polymerase chain reaction (PCR) is important in the application of genetics. It is used to produce billions of copies of a sample of DNA for use in processes such as genetic fingerprinting. PCR uses DNA polymerase enzymes that have a high optimum temperature, such as Taq polymerase. The process of PCR is as follows:

- The target DNA (the DNA to be copied) is mixed with DNA polymerase, nucleotides and primers. Primers are short sections of DNA that give a start point for the DNA polymerase to carry out DNA replication.

Knowledge check 47

In PCR, what temperature is used to break the hydrogen bonds and separate the two strands of DNA?

Examiner tip

Questions on PCR often focus on the different temperatures used. Make sure you know what they are and can explain why each temperature is used.

- The solution is heated to 95°C. This breaks the hydrogen bonds holding the two strands of the DNA together. The DNA is now single stranded.
- The solution is cooled to 55°C. This triggers the primers to join to their complementary bases on the target DNA.
- The solution is heated up to 70°C, which is the optimum temperature for the DNA polymerase. Using the primers as a starting point, the enzymes catalyse the formation of complementary DNA strands from the free nucleotides for both strands of the target DNA molecule.
- The above process is then repeated many times to produce billions of copies of the target DNA.

The polymerase chain reaction is shown in Figure 23.

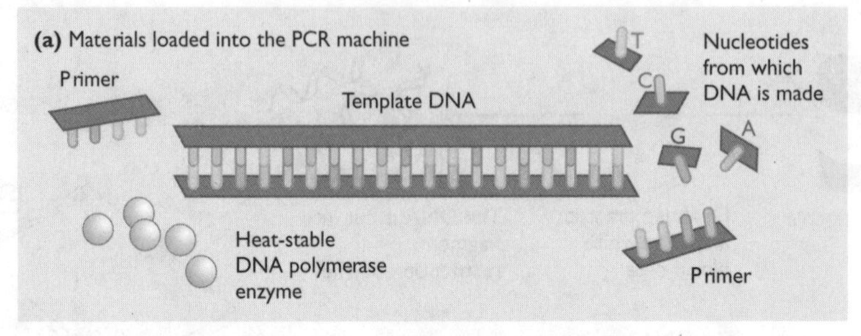

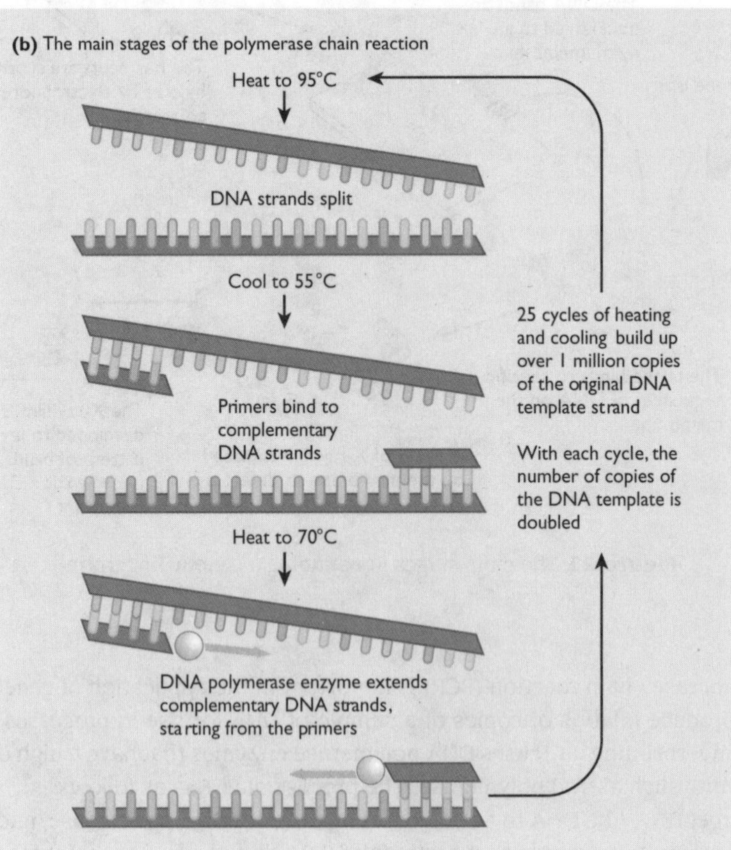

Figure 23 The polymerase chain reaction

Privacy issues

There are many issues raised by the ownership of genetic information and its misuse. As we become more able to predict conditions using genetic information it could be used to discriminate against people. As we become more reliant on genetic techniques for forensic investigation and the police hold more of our genetic information, there are increasing risks of false positives and possible miscarriages of justice.

Summary

- Cloning can be carried out by embryo cloning, somatic cell cloning or micropropagation of plants.
- Gene therapy involves the replacement of faulty genes that are causing a genetic disease with healthy genes. Vectors for gene therapy include liposomes that contain plasmids, and viruses.
- PCR can be used to produce many copies of a DNA molecule.
- Genetic fingerprinting can be used to compare samples of DNA. The technique is used in paternity testing and criminal investigations.
- Bacteria can be genetically modified using recombinant DNA technology to produce human proteins such as insulin.
- Crops can be genetically modified to give them desirable characteristics such as disease resistance or higher yield. Examples of genetically modified crops include soya and tomato.
- Genetic engineering raises ethical issues.

Energy and ecosystems

This is a short topic and the key to doing well is being able to apply the appropriate keywords in an answer.

- An **ecosystem** is all the **biotic** (living) and **abiotic** (non-living) factors in an area.
- **Population** — a group of organisms of the same species living in a particular area and interbreeding.
- **Community** — the different populations living in a particular area.
- **Niche** — an organism's niche is its place within an ecosystem, that is, its habitat, how it feeds, its response to predation and so on. No two species can have the same niche; interspecific competition would lead to one species outcompeting the other.
- **Microhabitat** — a small specialised habitat within a larger habitat.
- All organisms must gain energy and matter:
  - **Autotrophs** use simple inorganic compounds such as carbon dioxide and water to produce complex organic molecules.
  - **Heterotrophs** consume complex organic molecules derived from other living organisms.

Examiner tip

Make sure that you know what all the keywords mean.

Knowledge check 48

Is climate an example of a biotic factor or an abiotic factor?

Food webs

A **food web** represents the flow of energy through an ecosystem. Food chains (see Figure 24) can be put together to form food webs that show the feeding relationships between all the organisms in the ecosystem.

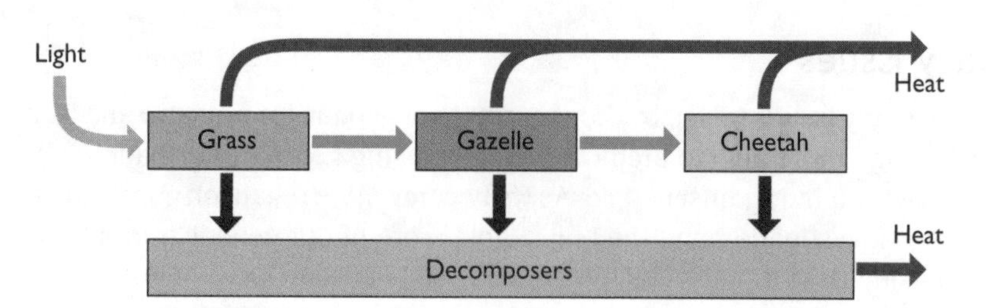

Figure 24 Flow of energy through a food chain

The source of energy for most ecosystems on Earth is light energy from the sun. In photosynthesis this light energy is converted into chemical energy. Organisms that do this are called **primary producers** and they are at the base of food chains. Examples of primary producers are autotrophic organisms such as algae and plants. All other organisms are consumers. Their classification depends on their position in the food chain:

- **Primary consumers** are animals that feed on the primary producers, i.e. they are **herbivores**.
- **Secondary consumers** are animals that feed on primary consumers, i.e. they are **carnivores**.

Knowledge check 49

What is the source of energy for most ecosystems?

Each new consumer occupies a higher trophic level on the food chain. The transfer of energy along a food chain is inefficient with much energy being lost through, for example, heat in respiration, inedible material such as bark or fur and through excretion. This is why food chains rarely have more than four or five levels.

Within food webs there are also **decomposers** (saprophytes), which feed extracellularly on dead organisms or non-living organic compounds. Decomposers have an important role in recycling nutrients in ecosystems.

Examples of two food webs are shown in Figure 25.

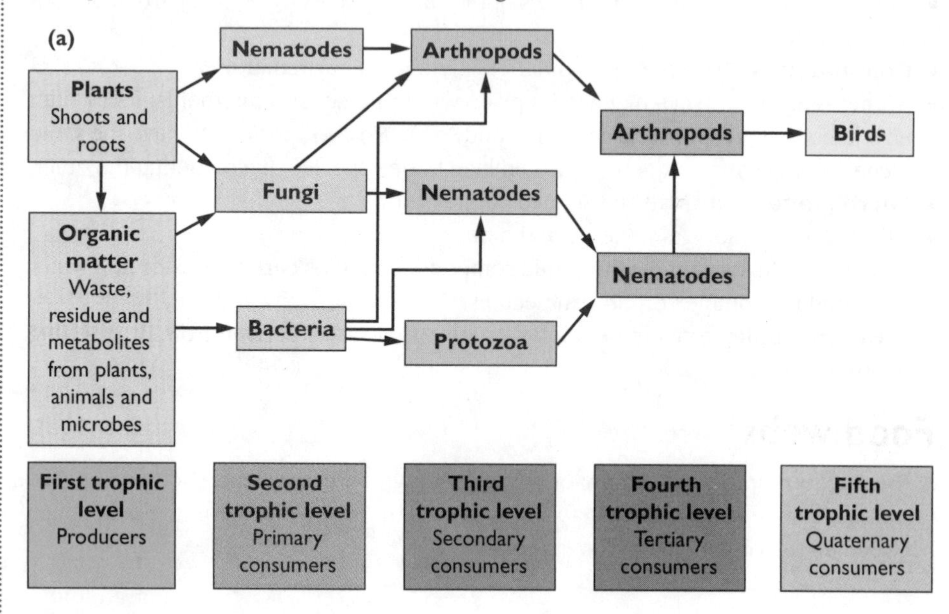

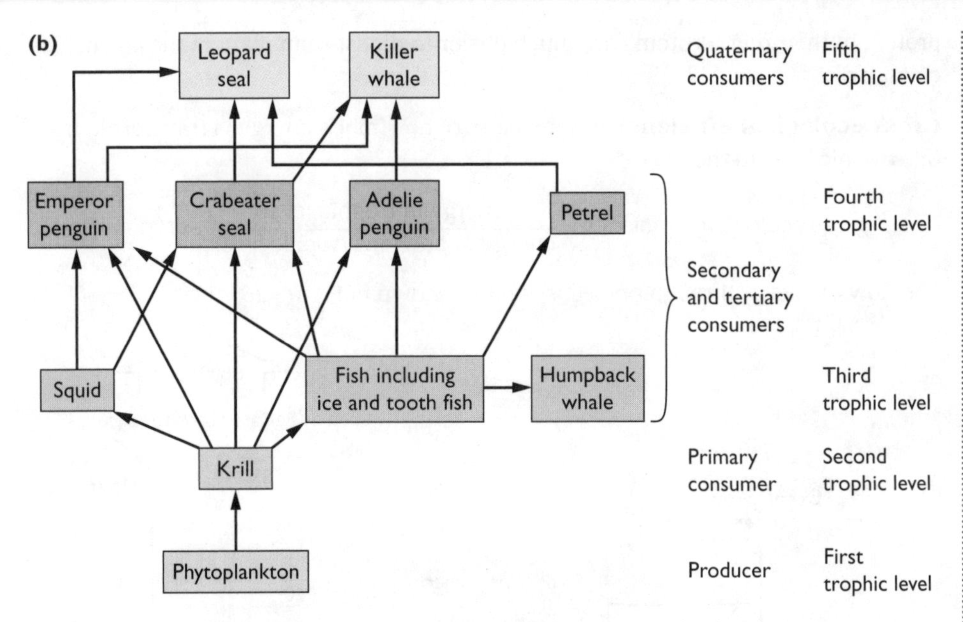

Figure 25 (a) A food web in soil; (b) A marine food web

Photosynthetic efficiency

Not all the light that falls on a leaf is utilised in photosynthesis. This may be due to:

- the light being the wrong wavelength to be absorbed by the photosynthetic pigments
- some of the light being reflected from, or transmitted through, the leaf and, therefore, not coming into contact with the pigments in the chloroplasts

The photosynthetic efficiency of a plant is a measure of how much of the light energy falling on a plant is incorporated into the biochemical products of photosynthesis. It is calculated as a percentage using the following equation:

$$\text{photosynthetic efficiency} = \frac{\text{energy incorporated into photosynthetic products}}{\text{total light energy falling on the plant}} \times 100\%$$

Plants rarely achieve photosynthetic efficiencies higher than 5%.

Gross primary productivity (GPP) is the rate at which producers make photosynthetic products.

A substantial amount of gross production is respired by the plant and therefore lost to the ecosystem. The production that is left over is the **net primary production (NPP).** The NPP represents the potential food available to primary consumers.

NPP = GPP – energy used in respiration by the plant

Herbivores have a lower efficiency of energy conversion compared to carnivores. This is because of the different diets of herbivores and carnivores. The diet of a herbivore contains a large amount of difficult to digest molecules such as cellulose. This means that herbivore digestion is relatively inefficient. In contrast to this carnivore digestion is relatively efficient due to the large amount of animal

Knowledge check 50

How does GPP differ from NPP?

Examiner tip

Make sure you are confident in calculating the flow of energy in an ecosystem and the various efficiencies. It is important to remember that energy cannot be created or destroyed, only transferred. This means that when energy flow is represented in a diagram the energy entering an organism must equal the energy leaving the organism and stored within it.

protein in their diet. Proteins are much easier to digest than plant material such as cellulose.

Gross ecological efficiency is a measure of how much energy is transferred from one trophic level to the next.

$$\text{gross ecological efficiency} = \frac{\text{energy in trophic level}}{\text{energy in previous trophic level}} \times 100$$

The flow of energy through one ecosystem is shown in Figure 26.

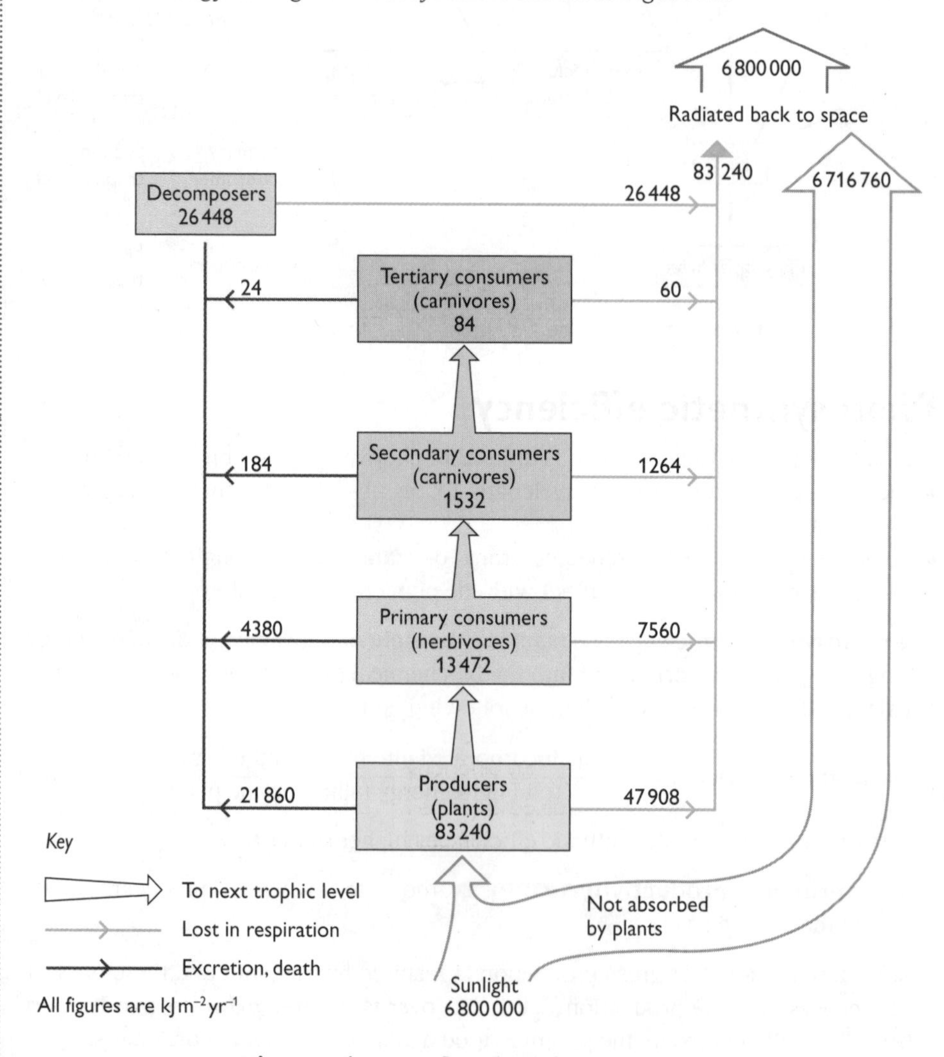

Figure 26 Energy flow through an ecosystem

Pyramids

Pyramids are used to represent food chains. A **pyramid of numbers** shows the number of organisms at each trophic level. Such data are relatively easy to collect, but drawing conclusions can be difficult, particularly as a pyramid of numbers can be inverted by large differences in size between organisms — for example a single plant might be fed upon by a large number of insects.

Knowledge check 51

Why might a pyramid of numbers be inverted?

A **pyramid of biomass** overcomes this problem by showing the biomass of the organisms at each trophic level. These data are more difficult to collect, but inversions are much rarer. They can occur due to temporal variations in populations. For example, after a phytoplankton (primary producer) bloom, reproduction by consumers (zooplankton) can lead to the zooplankton having a greater biomass than the phytoplankton for a short time before the death rate of the zooplankton increases and their total biomass falls.

All these problems are overcome by the use of a **pyramid of energy**. A pyramid of energy shows the energy in each trophic level per unit area, per unit time. This means units used in pyramids of energy must have an energy component, an area component and a time component. An example of the units used is $kJ\,m^{-2}\,year^{-1}$. As energy is lost at each trophic level from, for example, heat in respiration and no new energy is being inputted into the system it is impossible for pyramids of energy to be inverted. This allows ecologists to compare how efficient the energy transfer is between trophic levels. The disadvantage of pyramids of energy is that collecting the data to produce them is extremely difficult.

Pyramids of number, biomass and energy for three food chains are shown in Figure 27.

Knowledge check 52

What units are used when calculating pyramids of energy?

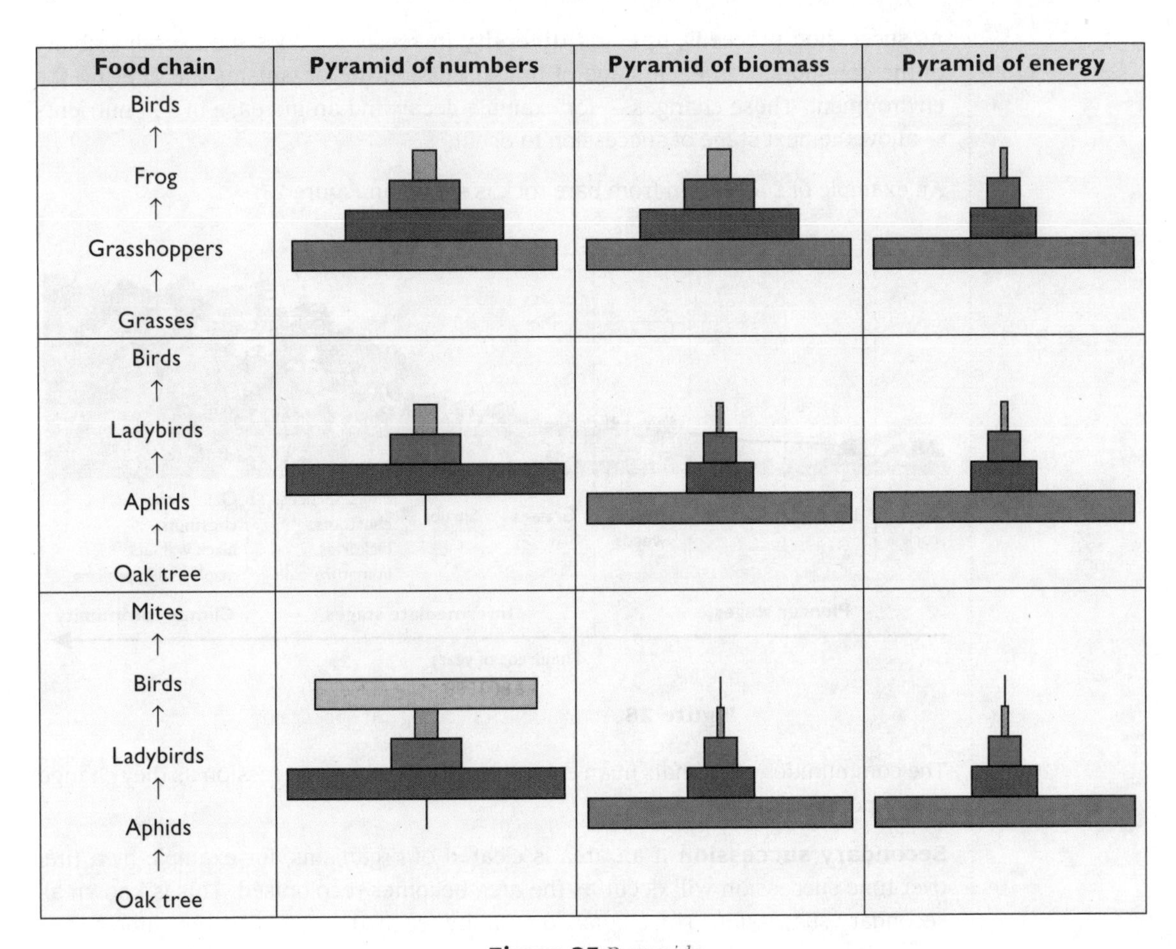

Figure 27 Pyramids

Succession

Succession is the change in the composition of the community of organisms in an ecosystem over time.

Primary succession refers to the introduction of plants/animals into areas that have not previously been colonised; **secondary succession** refers to the reintroduction of organisms into a bare habitat previously occupied by plants and animals.

Succession proceeds in a series of stages known as **serės**. Primary succession from bare rock occurs as follows:

- Algae and lichens colonise the bare rock. These organisms are called **pioneer species**.
- Erosion, and death and decay of the pioneer organisms leads to the development of simple soil and provides suitable conditions for mosses to grow.
- Further decay improves the soil to the point where grasses are able to grow.
- After grasses, shrubs will appear and then, after a long period of time, trees.
- Once the trees have formed a wood in an area, this is known as a **climax community** and is the stable end of succession.

Knowledge check 53

What is a sere?

As succession proceeds, **species diversity** increases as does the overall stability of the community. In each stage of the succession the organisms are altering the environment. These changes — for example decay and an increase in soil nutrients — allow the next stage of succession to occur.

An example of succession from bare rock is shown in Figure 28.

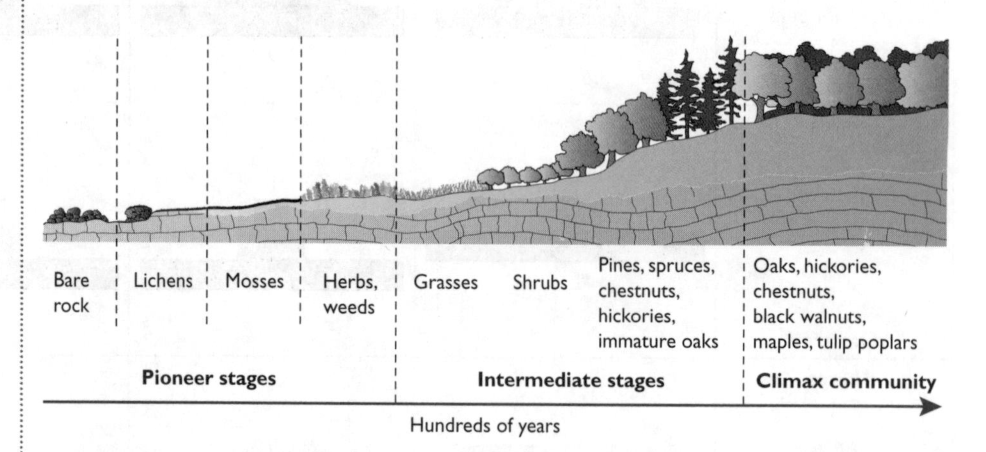

Figure 28 An example of ecological succession

The communities of animals in an environment also show succession as they change over a period of time.

Secondary succession If an area is cleared of organisms, for example by a fire, over time succession will occur as the area becomes recolonised. This is known as secondary succession. The climax community reached may not be the same as the one present before the destructive event.

Summary

- Energy is transferred in ecosystems. The source of energy for most ecosystems on Earth is the sun. Primary producers use light energy in photosynthesis to produce complex organic molecules. Primary consumers feed on producers and are themselves fed on by secondary consumers. Decomposers feed on dead organisms and non-living organic compounds.
- Energy is lost at each trophic level, mainly as heat from respiration. Because of their differing diets primary consumers are less efficient energy converters than secondary consumers.
- Plants do not use all the light energy that falls on them to form photosynthetic products. Their efficiency is known as photosynthetic efficiency.
- The gross primary productivity (GPP) is a measure of the formation of organic compounds by plants. The net primary productivity is GPP minus the products used in respiration. NPP is what is available to the primary consumers in a food web.
- The gross ecological efficiency is a measure of the efficiency of energy transfer through trophic levels in a food web.
- Pyramids of energy can be used to represent energy flow through ecosystems.
- Succession is the change in the species composition of an ecosystem over time. Primary succession occurs in areas where life has not existed previously. Secondary succession occurs in areas that have had a community of organisms which has been removed.

The effects of human activities and sustainability

This topic deals with the environmental problems caused by human activity. Humans can have a number of detrimental effects on ecosystems. These include deforestation, overfishing and overexploitation of agriculture, all of which can lead to a reduction in biodiversity. There are a number of ways the damage from these can be alleviated and a number of different conservation methods that can be used to preserve species.

Production of carbon dioxide by human activities and the reduction of photosynthesis because of deforestation are increasing atmospheric carbon dioxide. This is leading to climate change, which has massive ecological implications all over the world.

Human effects on selection

Humans have been intentionally affecting selection by **selective breeding** for a very long time. This is known as **artificial selection**. Humans breed animals and plants that have desirable characteristics in order to produce offspring that have the same desirable characteristics. This process takes a long time, over many generations. It has given rise to many organisms that humans use. It also leads to the same species having different breeds or varieties with different characteristics. A good example of this is the domestic dog. Dogs are all the same species (*Canis lupus familiaris*) but have been bred selectively to produce different breeds with different characteristics — for example German shepherd and chihuahua.

Selectively breeding cows for milk production is outlined below:

- A cow that produces a large volume of milk is bred with a bull whose mother produced a large volume of milk.

- Some of the offspring will be female and will produce large volumes of milk.
- These cows can then be bred with other bulls whose mothers produced large volumes of milk.
- This continues over many generations to give cows that produce large volumes of milk.

However humans have also inadvertently influenced natural selection. Two examples of this you need to study are warfarin resistance in rats and antibiotic resistance in bacteria.

Warfarin resistance in rats

The anticoagulant warfarin has been used successfully as a rat poison. However, it has been found that some rats have a mutation that makes them immune to warfarin. A codominant allele, R, gives resistance to warfarin and confers a high demand for vitamin K. The other codominant allele, N, does not give warfarin resistance and does not confer the high vitamin K demand. In natural conditions, the vitamin K demand would be a selective disadvantage. Animals that are homozygous for the resistance allele (RR) have a high vitamin K demand and are immune to warfarin. Heterozygous (RN) rats have a lower vitamin K demand while still being immune to warfarin. In areas where warfarin is used, the R allele is a selective advantage because it prevents rats from being killed by warfarin. Therefore, the frequency of this allele in the population increases. The heterozygote (RN) is the most advantageous combination of alleles because it confers warfarin resistance without the high vitamin K demand of the homozygote. However if the heterozygote is favoured, the high frequency of the dominant allele in the gene pool will also lead to homozygous dominant rats.

Knowledge check 54

In what conditions would warfarin resistance become a selective advantage to rats?

Examiner tip

You have to learn these two examples of unintentional selection by humans. In the exam you may be given other examples and have to apply your knowledge. The key thing is not to panic! Read the example carefully and try and think about how your knowledge could be applied to the question.

Antibiotic resistance in pathogenic bacteria

Since the first antibiotics were discovered they have been used widely, not just in human medicine but also in agriculture where their use has been a great advantage, particularly in the intensive farming of animals. However, the high use of antibiotics has created a selection pressure that has given bacteria that have mutated to become resistant to antibiotics a great selective advantage. MRSA is an example of bacteria that are resistant to many types of antibiotics.

Endangered species and conservation

Many species around the world are endangered and at threat of becoming extinct. Some of the reasons for this include:

- **Habitat destruction** — for example deforestation, drainage of wetland areas and the conversion of land to agriculture or for building
- **Hunting and overexploitation** — this can range from hunting large mammals for their fur or tusks to overfishing of natural populations of marine fish
- **Pollution** — polluting a habitat kills or displaces wildlife
- **Competition** from species introduced to an area by human activity — an example is competition from grey squirrels imported from the USA causing near extinction of indigenous British red squirrels
- **Climate change** — alters ecosystems, making it difficult for some organisms to survive

Through **conservation** humans can work to preserve biodiversity, ensure the survival of endangered species and conserve natural gene pools.

Conservation is the maintenance of the biosphere and enhancement of biodiversity. It can also be defined as the planned preservation of wildlife. There are a number of different strategies that are used in an attempt to preserve wildlife:

- Rare breed societies can maintain populations of varieties of animals that are more traditional and may no longer be popular or commercially viable. This ensures the preservation of their unique alleles, preventing them from being lost from the gene pool. An example of a rare breed of dog that could be conserved in this way is the otterhound.
- Organisms can be reintroduced to areas where they have become locally extinct. A successful example of this is the reintroduction of red kites to Wales.
- Seed banks, such as the one at the Royal Botanical Gardens at Kew, are used to preserve seeds of varieties or species of plants that are on the brink of extinction.
- Sperm or embryo banks can be established. These could be used in breeding programmes of animals that are near extinction.
- Zoos can carry out captive breeding programmes of animals that are near extinct or are extinct in the wild. Cheetahs are being successfully bred in captivity at the moment.
- International cooperation and treaties can also be used. Examples include bans on commercial whaling and the trade in ivory.
- Habitats, for example wetlands and coral reefs can be protected from development.
- Legislation (laws) can be used to ban activities that are detrimental to conservation, for example making it illegal to collect birds' eggs and banning the picking of wild flowers.
- Ecotourism is a sustainable method of tourism that aims to conserve species and habitats while also providing income from tourism for local people.

There are also specific conservation strategies for some of the environmental problems which are discussed later.

Knowledge check 55

What is conservation?

Examiner tip

When answering questions on conservation, be specific. It is easy to write a lot of general information about, for example, pandas without earning any marks. Examples are fine, but you must focus on the conservation strategies in the BY5 specification.

Agricultural exploitation

As the world's population increases and becomes richer, the pressure on agricultural food production has increased. This has led to:

- **Increased mechanisation** More and more machines are used in farming. This has necessitated the use of bigger fields, so hedgerows (important habitats) have been removed. This has reduced biodiversity.
- **Monoculture** This is the growth of a single crop. This creates problems in terms of the crop constantly taking the same nutrients out of the soil, which are then not replaced. It also has caused an increase in pests and disease because of the densely planted crops. This has led to an increase in the use of pesticides.

In recent years there has been an emphasis on more sustainable agriculture and on the role farmers can play in maintaining and increasing biodiversity. Financial incentives can be given to farmers to encourage them to increase biodiversity and to promote conservation.

Water availability will be a big issue in the future and drought-resistant crops will have to be developed in order to maintain food production.

Deforestation

Deforestation is the permanent removal of forest. On a global scale, deforestation is being driven by factors such as the need to clear forests to provide agricultural land for subsistence farming, cash crops or the extraction of timber. Some of the environmental implications of deforestation include:

- reduction in biodiversity due to loss of habitat
- loss of beneficial plants — plants that could provide medical benefits to humans may become extinct
- soil erosion — caused by exposure of the soil to the wind and the removal of tree roots
- changes in local rainfall patterns and flooding
- climate change — on a global scale removal of trees reduces the fixing of atmospheric carbon dioxide for photosynthesis; high carbon dioxide levels in the atmosphere drive climate change

Knowledge check 56

Why is deforestation a contributory factor to climate change?

In order to prevent deforestation but still ensure that timber can be produced forests must be managed sustainably. The following processes allow forests to be managed sustainably:

- **Coppicing** In coppicing, trees are cut down to the trunk. Many trees then re-grow from the trunk. This means timber is still produced, but the tree is not killed and re-grows to produce more timber.
- **Selective cutting** This involves cutting down the oldest, largest trees, allowing the smaller, younger trees to grow. It can also be used to minimise effects such as soil erosion by leaving some trees to hold the soil in place.
- **Planting trees optimum distances apart** reduces intraspecific competition, therefore encouraging faster growth and reducing the instance of disease.
- **Planting fast-growing trees** and having slower rotation times can also help the sustainable management of forests.

An important part of forestry conservation is the preservation of native woodland, i.e. woodland that has developed naturally and has not been planted to provide timber. By preserving native woodland, local biodiversity is enhanced.

Overfishing

Overfishing is a huge problem around the world. Fish populations are falling due to intensive, mechanised fishing. There are also knock-on effects for other organisms through the disruption of food webs by the removal of large numbers of a fish species. There are steps that can be taken to reduce the impact of overfishing:

- **Increase mesh size of nets** — this allows smaller, younger fish to escape the nets and go on to reproduce
- **Protected areas** — fishing can be banned in a certain area; this is especially important in areas where fish reproduce
- **Closed season** — fishing is banned for part of the year, timed to coincide with when fish reproduce
- **Quotas** — fishermen are only allowed to land a certain number of fish

Examiner tip

As with conservation, when answering questions on deforestation or overfishing make sure that you write detailed answers that use concepts from the specification. Do not just rely on your general knowledge.

The farming of fish also reduces the pressure on wild stocks. However fish farming has potential disadvantages. It is an intensive farming process that leads to a high incidence of disease and pests. To combat these problems farmers have to use lots of antibiotics and pesticides. Runoff from fish farms can also lead to eutrophication in local bodies of water.

Knowledge check 57

How can changing the mesh size of nets help to reduce overfishing?

The carbon cycle

Figure 29 shows the **carbon cycle**, which should be familiar from BY4.

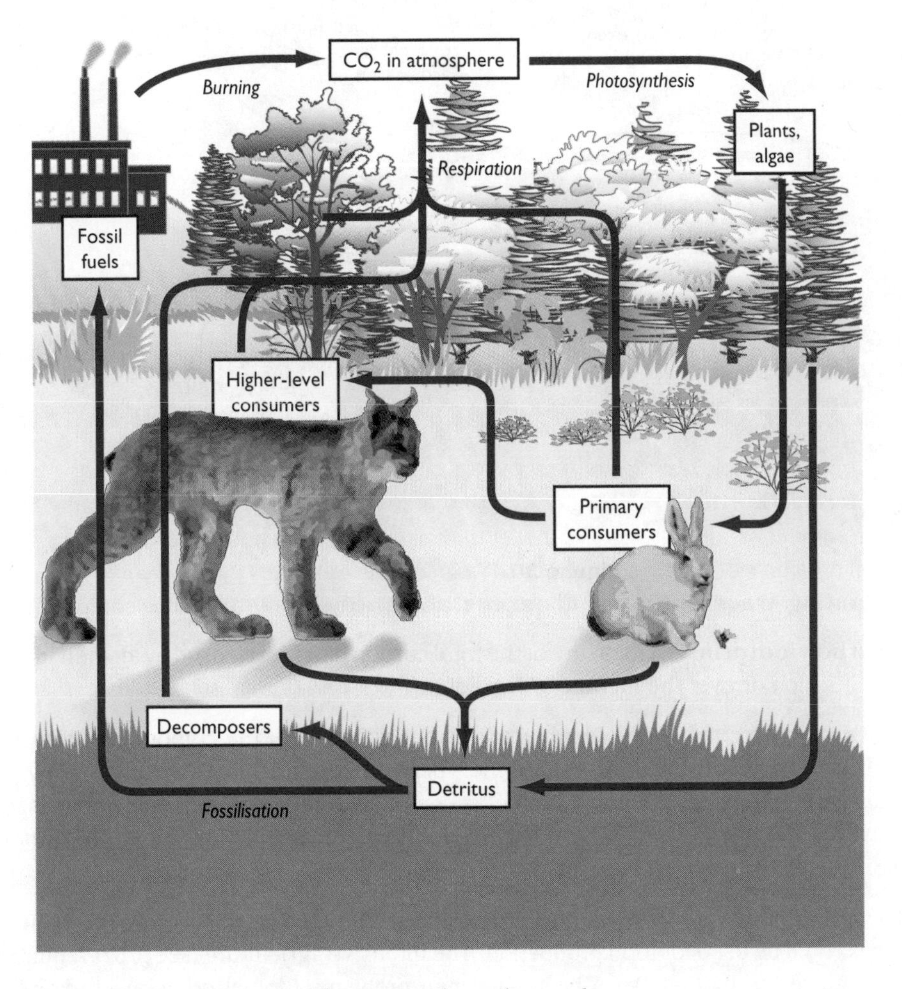

Figure 29 The carbon cycle

The carbon cycle can be altered by human activity, for example by burning fossil fuels (releasing carbon dioxide) and cutting down trees (reducing the fixing of atmospheric carbon dioxide). An increase in atmospheric carbon dioxide contributes to the **greenhouse effect** (Figure 30), which is one of the main driving forces of climate change.

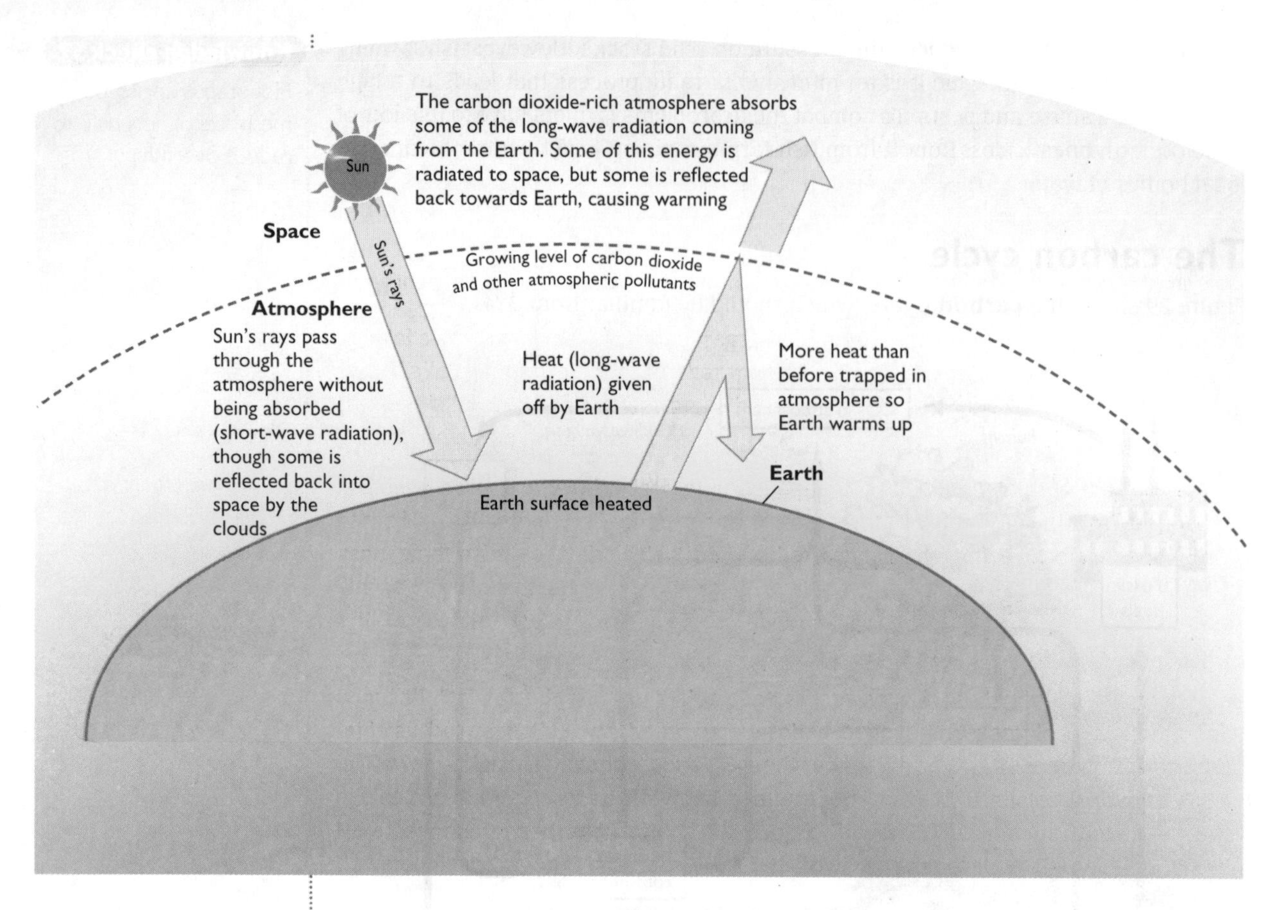

Figure 30 The greenhouse effect

A **carbon footprint** is a measure of the total carbon dioxide released by an individual, organisation or over the lifetime of a product.

One strategy for reducing greenhouse gas emissions from the combustion of fossil fuels is to use **biofuels**. Biofuels are fuels derived from plants. As the plant grows it fixes carbon dioxide. When the biofuel is combusted the carbon dioxide is released. However, the plants being grown to provide the biofuels will continue to fix carbon dioxide and the cycle will continue.

There are, however, disadvantages to biofuels. The land used to grow the biofuel crops could be used for food production. The intensive agriculture (see p. 57) required to produce the biofuel crops brings its own problems Biofuel crops are grown as monocultures (see p. 57), with all the associated problems that reduction in biodiversity brings.

Knowledge check 58

What are biofuels?

Fertiliser use and eutrophication

Nitrate level is often the limiting factor for the growth of plants. To increase crop yield nitrate-containing fertilisers are used and this has potential environmental consequences. One of these is **eutrophication** in local bodies of water such as ponds or lakes. The process of eutrophication is as follows:

- Eutrophication occurs when fertilisers or sewage enter a body of water. This can be through runoff from the surface or through leaching through the soil.
- Bodies of water contain populations of single-celled algae called phytoplankton. Phytoplankton growth is normally nutrient limited, particularly by a lack of nitrates.
- The nitrates in the fertiliser remove this limitation on phytoplankton growth and there is a large and rapid growth in the phytoplankton population. This is known as a phytoplankton bloom. A phytoplankton bloom is damaging in two ways:
  - The large number of cells forms a 'skin' on the water that blocks the light from the algae and plants that live at the bottom of the lake or pond.
  - The algae and plants cannot photosynthesise, so they die.

  The phytoplankton bloom is relatively short-lived and soon the single-celled algae that make up the bloom die.
- As the plants, algae and phytoplankton die they are decomposed by bacteria. These bacteria have a high biochemical oxygen demand (BOD) due to taking in oxygen as they carry out decomposition. This leads to a fall in the oxygen concentration in the water.
- This fall in oxygen concentration in the water leads to the death of other aquatic organisms, which themselves are decomposed by bacteria. This removes more oxygen from the water, which can lead to the water becoming anoxic (without oxygen). The result is a huge reduction in the biodiversity of aquatic organisms in the lake or pond.

Knowledge check 59

What does BOD stand for?

Summary

- Through selective breeding, humans can artificially select plant or animal characteristics that are desirable. Human activity has also inadvertently affected selection. Examples include warfarin resistance in rats and antibiotic resistance in bacteria.
- Many species have become endangered through human activity. Conservation is the planned preservation of wildlife and there are many conservation strategies that can be used to preserve biodiversity.
- Agriculture can have a detrimental effect on biodiversity particularly through increased mechanisation and monoculture.
- Deforestation and overfishing are both serious issues affecting the natural world. There are a number of different ways of reducing the harm they cause.
- Human activity such as burning fossil fuels and deforestation is leading to an increase in the concentration of carbon dioxide in the atmosphere. This is a factor in climate change.
- Biofuels are potentially sustainable forms of fuel. However, their production has disadvantages, particularly the use of land that could be used to grow food crops.
- Fertilisers entering bodies of water can lead to eutrophication.

Questions & Answers

This section contains questions on each of the topic areas in the specification. They are written in the same style as the questions in the BY5 exam so they will give you an idea of the sort of thing you will be asked to do in the exam. After each question there are answers by two different students followed by examiner's comments on what they have written. These are important because they give you an insight into the responses the examiners are looking for in the exam. They also highlight some of the common mistakes students make.

The A2 Unit 5 paper

The BY5 examination lasts 1 hour 45 minutes and is worth 80 marks. The first 70 marks are for answering structured questions and the last 10 marks are for writing an essay. In the exam you will be given a choice of two essay questions but you only need to answer one of them.

In addition to sample structured questions, this section of the guide also contains an example essay question.

Examiner's comments

Examiner comments on the questions are preceded by the icon e. They offer tips on what you need to do in order to gain full marks. All student responses are followed by examiner's comments, indicated by the icon e, which highlight where credit is due. In the weaker answers, they also point out areas for improvement, specific problems and common errors such as lack of clarity, irrelevance, misinterpretation of the question and mistaken meanings of terms.

Question 1 DNA replication

(a) In what stage of the cell cycle does DNA replication occur? (1 mark)

Meselson and Stahl's investigation into DNA replication relied on DNA nucleotides containing different isotopes of nitrogen, ^{14}N and ^{15}N.

(b) What part of the DNA nucleotide contained ^{14}N or ^{15}N? (1 mark)

Meselson and Stahl cultured bacteria in a medium containing ^{15}N. After cell lysis and centrifugation, the position of the DNA in the tube was as shown in the diagram below. They then transferred the bacteria to a medium containing ^{14}N. After each division of bacteria they centrifuged DNA sampled from some of the cells.

(c) Complete the diagrams below to show the positions of DNA in the tubes for the second and third generations. (3 marks)

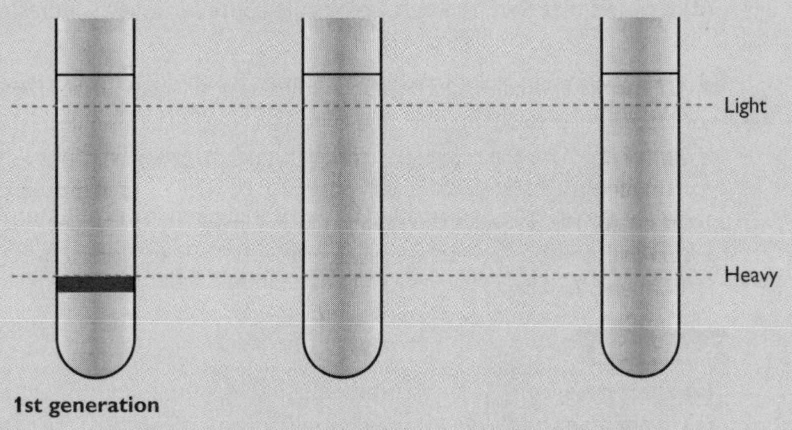

(d) Explain how this provided evidence of semi-conservative replication. (2 marks)

ⓔ This question requires a small amount of synoptic knowledge from AS. It is important that you are comfortable with key concepts such as cell division, biochemistry, enzymes and transport across membranes. Most marks in this question are related to Meselson and Stahl's investigation into DNA replication. Make sure that you understand this experiment fully and can both recall and explain the results.

Student A

(a) In prophase of mitosis
(b) The nitrogenous base

(c)

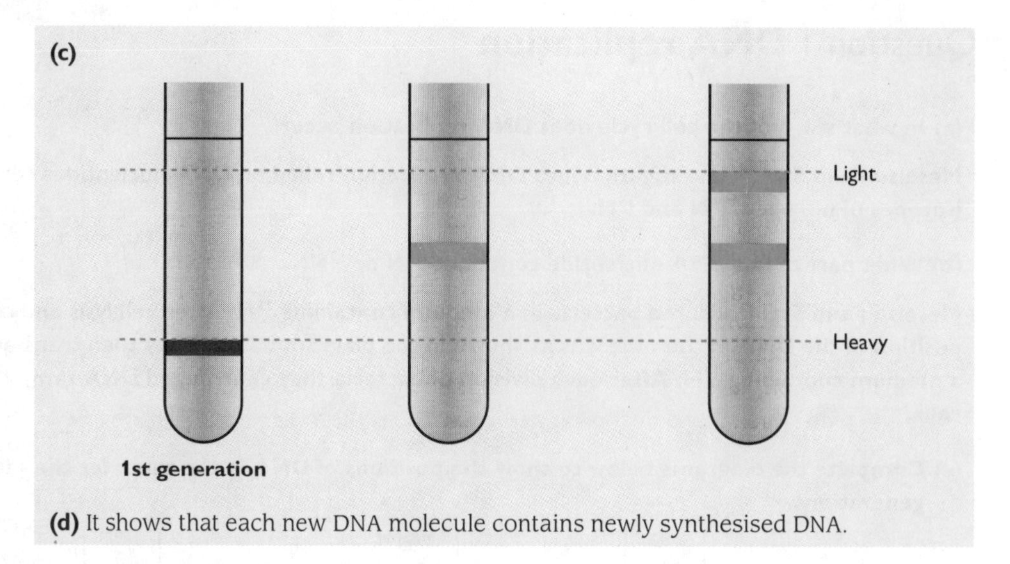

(d) It shows that each new DNA molecule contains newly synthesised DNA.

e 4/7 marks awarded (a) This is an example of a synoptic question. Synoptic questions test knowledge from previous modules. It is important that you have an understanding of some key AS concepts. In this case, the student has made a common mistake in thinking that DNA replication occurs in mitosis. DNA replication is important in both mitosis and meiosis and occurs during interphase. (b) This answer is correct and scores 1 mark. (c) The student has drawn the correct bands in both tubes so scores the available marks. (d) This answer is too simplistic to score either of the 2 marks available.

Student B

(a) Interphase
(b) In adenine
(c)

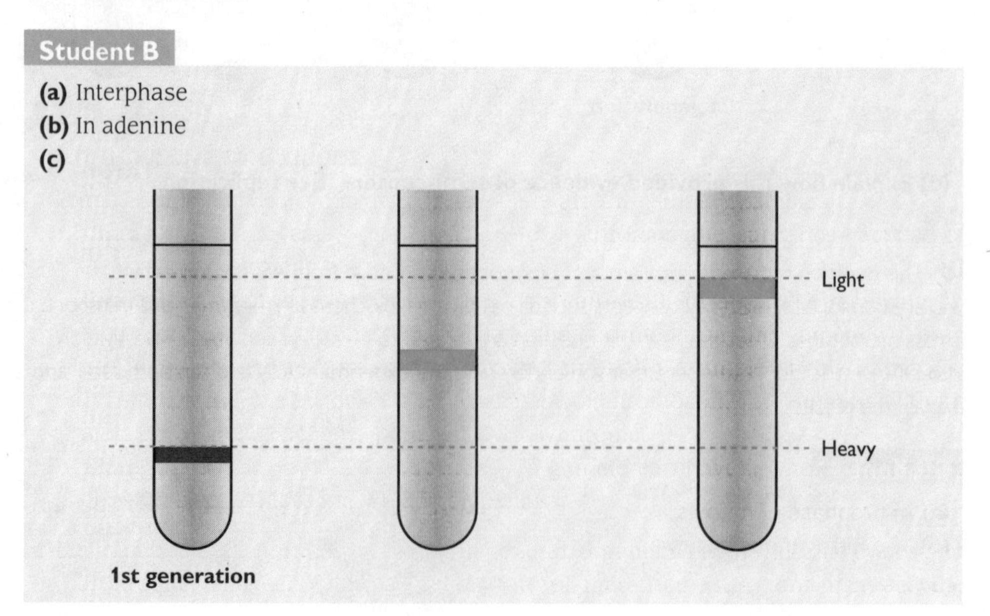

(d) This investigation showed DNA was semi-conservatively replicated as it showed that each DNA molecule contained one original strand and one newly synthesised strand. This was by shown by the continued presence of the ^{15}N-strand (forming the hybrid $^{15}N/^{14}N$ strand) and the formation of a ^{14}N-strand in the third generation.

e 5/7 marks awarded (a) This correct answer scores 1 mark. (b) This answer is too specific. Adenine is a nitrogenous base and would contain ^{14}N or ^{15}N, but so would the other three nitrogenous bases. (c) The student has only drawn two bands on the tubes. Since this is a 3-mark question, the question is guiding you to draw three bands. The two drawn are correct but the missing third band means that the student only scores 2 marks. (d) This excellent answer scores both marks. There is 1 mark for describing what semi-conservative replication is (one original and one newly-formed strand) and 1 mark for relating the results to this (the continued presence of the ^{15}N strand and the formation of DNA molecules containing only ^{14}N).

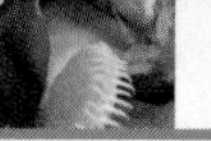

Question 2 Protein synthesis

Below is a sequence of mRNA:

UGCGAGGGG

(a) Write out the sequence of bases in the DNA strand that this mRNA strand was produced from. (1 mark)

(b) Name the enzyme that catalyses the production of mRNA. (1 mark)

(c) Using the table below, determine the sequence of amino acids coded for by this strand of mRNA. (1 mark)

The results of two mutations in the DNA that coded for the mRNA sequence are shown below:

Mutation 1 — UGCGAGGGA

Mutation 2 — UGCGAGGAG

(d) Using the codon table, explain why mutation 2 might have much more serious consequences than mutation 1. (4 marks)

Second position

First position (5′ end)	U	C	A	G	Third position (3′ end)
U	UUU Phe	UCU Ser	UAU Tyr	UGU Cys	U
	UUC Phe	UCC Ser	UAC Tyr	UGC Cys	C
	UUA Leu	UCA Ser	UAA stop	UGA stop	A
	UUG Leu	UCG Ser	UAG stop	UGG Trp	G
C	CUU Leu	CCU Pro	CAU His	CGU Arg	U
	CUC Leu	CCC Pro	CAC His	CGC Arg	C
	CUA Leu	CCA Pro	CAA Gln	CGA Arg	A
	CUG Leu	CCG Pro	CAG Gln	CGG Arg	G
A	AUU Ile	ACU Thr	AAU Asn	AGU Ser	U
	AUC Ile	ACC Thr	AAC Asn	AGC Ser	C
	AUA Ile	ACA Thr	AAA Lys	AGA Arg	A
	AUG Met	ACG Thr	AAG Lys	AGG Arg	G
G	GUU Val	GCU Ala	GAU Asp	GGU Gly	U
	GUC Val	GCC Ala	GAC Asp	GGC Gly	C
	GUA Val	GCA Ala	GAA Glu	GGA Gly	A
	GUG Val	GCG Ala	GAG Glu	GGG Gly	G

e This is a straightforward question. However, it is important in any question that involves genetic codes to be careful when writing out the answers and to check them thoroughly. The final part is a 4-mark question. If a question has this many marks, make sure that you answer it in sufficient detail.

Student A

(a) ACGCUCCCC
(b) RNA polymerase
(c) Cysteine, glutamic acid, serine
(d) Mutation 2 appears to be a much more serious mutation than mutation 1. It could lead to the incorrect formation of the polypeptide because the primary sequence of the polypeptide (the sequence of the amino acids) will be different. This could lead to the protein being non-functional.

e 4/7 marks awarded (a) The student has made a common mistake when writing out the complementary sequence — failing to replace U (uracil) with T (thymine). Remember that uracil replaces thymine in RNA, but it is thymine that occurs in DNA. This is an example of a simple mistake that has cost the student an easy mark. It is so important to check over your answers to avoid mistakes like this. (b) This is the correct answer and scores 1 mark. (c) The student has made a simple mistake and lost an easy mark. (d) This is a good answer. Although the student has not realised why mutation 2 is more serious than mutation 1, the explanation of the possible consequences of DNA mutation is good and scores 3 of the 4 marks. The fourth mark is for saying that mutation 1 is less serious as the same amino acid (glycine) will be coded for by the final codon (both GGG and GGA code for glycine).

Student B

(a) ACGCTCCCC
(b) RNA polymerase
(c) Cysteine, glutamic acid, glycine
(d) As mutation 1 is a substitution mutation, the replacement of the final G in the last codon with A means that glycine is still coded for and so the correct amino acid is still produced.

e 4/7 marks awarded (a) This is the correct DNA sequence and scores the 1 mark available. (b) RNA polymerase is the correct enzyme. (c) This is the correct sequence of amino acids so scores the 1 mark available. (d) The student understands the idea behind this question but the answer is not developed. The student scores 1 mark for correctly stating that in mutation 1 the correct amino acid is still produced. However, there is no explanation of why mutation 2 is potentially more serious. Because an incorrect amino acid is produced, the primary structure is altered, which could result in the protein being non-functional. This question illustrates how important it is to look carefully at the mark allocation for each part-question to ensure that you include enough detail in your answers to score all the marks available.

Question 3 **Sexual reproduction in humans**

Complete the gaps in the paragraph below on fertilisation. (7 marks)

During sexual intercourse sperm are deposited at the top of the vagina. They swim up through the cervix and the uterus to the ....(a)............................ where they will meet the oocyte. Before fertilisation can take place the sperm must go undergo ....(b)............................ which is a change in the membrane around the acrosome. When the sperm comes into contact with the oocyte the ....(c)............................ reaction occurs. This leads to a release of enzymes which digest through the corona radiate and the ....(d)............................ To prevent any further sperm entering the oocyte a ....(e)............................ forms. The entry of the sperm causes the oocyte to complete ....(f)............................ and fertilisation occurs when the two nuclei fuse, forming a ....(g)............................

'Fill in the blanks' questions can be a source of easy marks. However, one mistake can lead to further incorrect answers, which can soon add up. The key to this type of question is to read the paragraph through carefully at least twice before putting any answers down. Be wary of basing an answer on a previous answer. If the first answer is wrong your reasoning to get the second answer will also be wrong and you will lose 2 marks.

Student A

(a) Uterus
(b) Capacitation
(c) Acrosome
(d) Zona pellucida
(e) Fertilisation membrane
(f) Mitosis
(g) Fetus

4/7 marks awarded (a) This is incorrect. The student has not read the question carefully. It is true that sperm swim up through the uterus, but this is mentioned in the question. The correct answer is oviduct. (b)–(e) These answers are all correct. (f) This is incorrect. The oocyte completes meiosis II, not mitosis. This highlights the importance of knowing the difference between key terms such as mitosis and meiosis. (g) This is also incorrect. The developing child is only known as a fetus 9 weeks after fertilisation has occurred. The immediate product of fertilisation is a zygote.

Student B

(a) Oviduct
(b) Acrosome
(c) Capacitation
(d) Zona pellucida
(e) Fertilisation membrane
(f) Meiosis II
(g) Zygote

5/7 marks awarded This student has correctly answered (a) and (d)–(f). However, capacitation and the acrosome reaction have been mixed up and so the student loses the 2 marks for (b) and (c). Make sure you learn thoroughly all the stages of processes such as fertilisation.

Question 4 Sexual reproduction in plants

(a) What is meant by cross-pollination? (2 marks)

(b) Give one advantage of cross-pollination over self-pollination. (1 mark)

(c) Describe three differences between insect-pollinated and wind-pollinated flowers. (3 marks)

e Questions on pollination are common and should be an easy source of marks. Make sure you know all the key terms. When describing differences make two points for each mark — for example 'Insect-pollinated flowers have… whereas wind-pollinated flowers have…'.

Student A

(a) Pollination is plant reproduction.
(b) Increases genetic variation
(c) Insect-pollinated flowers have large brightly coloured petals whereas wind-pollinated flowers have flowers that are not coloured. The anthers of insect-pollinated flowers are inside the flower whereas in wind-pollinated flowers the anthers hang outside the flower. Insect-pollinated flowers produce sticky pollen whereas wind-pollinated flowers produce large quantities of pollen.

e **2/6 marks awarded** (a) Pollination is involved in plant reproduction but an overly simple answer such as this will not score any marks. It is important to learn definitions of key terms such as pollination because questions like this then become a source of easy marks. (b) This is correct and scores 1 mark. (c) The correct word used to describe the colour of the petals in wind-pollinated plants is 'dull'. 'Not coloured' is not a good enough answer to score a mark here. The point made about anthers is correct and scores a mark. The third sentence falls into a common trap. Both the points made are correct but they are not a direct comparison. A statement such as 'insect-pollinated flowers produce sticky pollen whereas wind- pollinated flowers produce smooth pollen' would score a mark. The student scores only 1 mark for this is relatively easy recall question.

Student B

(a) Cross-pollination is the transfer of a pollen grain from the anther of one plant to the stigma of another plant of the same species.
(b) Cross-pollination reduces the chance of recessive alleles causing genetic diseases.
(c) A wind-pollinated flower will have a large feathery stigma whereas an insect-pollinated flower will have smaller, non-feathery stigmas. Insect-pollinated flowers produce scent to attract insects whereas wind-pollinated flowers are not scented. Wind-pollinated flowers produce large volumes of light pollen whereas insect-pollinated flowers produce smaller volumes of sticky pollen.

e **6/6 marks awarded** (a) This is a good answer and scores both marks. There is 1 mark for the general description of pollination and 1 mark for explaining 'cross'. (b) This is correct, for 1 mark. (c) This is an excellent answer. Three clear differences are outlined and the student scores all 3 marks.

Question 5 Inheritance

(a) In a breed of cats, the allele for red fur is dominant to the allele for white fur. The allele for thick fur is dominant to the allele for thinner fur. A cat with white thin fur was bred with a homozygous cat with thick red fur. Using a genetic cross, find the expected ratios of the genotypes and phenotypes of the offspring. (4 marks)

(b) One of the offspring cats was bred with a cat that had white, thick fur; the mother of this cat had thin fur. Is it possible for these two cats to have offspring with thin fur? If so, what is the chance of this occurring? Show your working. (5 marks)

e This question involves a dihybrid cross. It is challenging because you have to choose the letters to use for the cross and work out a genotype based on an organism's parent. The key to genetic crosses is to take your time and work logically through the problem. Check your work several times to ensure that errors do not creep in.

Student A

(a) Parental genotypes: RRTT rrtt

Gametes: (RT) (rt)

Offspring:

	(rt)
(RT)	RrTt

Offspring phenotype: 100% RrTt

Offspring genotype: 100% of cats would have red, thick fur

(b) As the parents both have the recessive allele for thin fur there is a chance of them having offspring with thin fur. Their offspring genotypes would be in the ratio 9:3:3:1, so therefore the chance of them having an offspring with thin fur would be 1 in 16.

e **3/9 marks awarded** (a) The student scores 1 mark for correctly stating the genotypes of the parents and the gametes produced and 1 mark for completing the cross correctly. However, 2 marks are lost by mixing up the terms phenotype and genotype. The genotype is the alleles in the organism (RrTt) and the phenotype is the physical characteristics of the organism (red, thick fur). Having done the hard work the student loses marks through a simple error. (b) The student has not felt able to answer this question and has guessed the answer. This question requires a genetic cross. The first step is to work out the parental genotypes. One parent is the offspring from (a) so the genotype is RrTt. The other parent has white, thick fur. This means that the genotype could be either rrTT or rrTt. This is where the next statement in the question helps — this cat's mother had thin fur. To have thin fur, her genotype must be tt. One of these alleles would be passed to her offspring, so the genotype of the cat in the question is rrTt. Once you know this, carry out the genetic cross to see if a thin-furred offspring is possible. Student A has been unable to follow this logic and has given the memorised dihybrid ratio and made the same mistake of mixing up the terms genotype and phenotype. The statement 'as both parents have the recessive allele for thin fur there is a chance of them having offspring with thin fur' earns 1 mark.

Student B

(a) Parental genotypes: RRTT rrtt
Gametes: (RT) (rt)
Offspring:

	(rt)
(RT)	RrTt

Offspring genotype: 100% RrTt
Offspring phenotype: 100% of cats would have red, thick fur.

(b) Parents: RrTt rrTt
Gametes: (RT) (rt) (Rt) (rT) (rT) (rt)
Offspring:

	(rT)	(rt)
(RT)	RrTT	RrTt
(Rt)	RrTt	Rrtt
(rT)	rrTT	rrTt
(rt)	rrTt	rrtt

Yes, it is possible. There is a 25% chance of an offspring having thin fur.

9/9 marks awarded (a) This is an excellent answer that scores all 4 marks. (b) This is also an excellent answer, which scores all the 5 marks available. The student has deduced correctly the genotypes of the parents and written out the gametes correctly. The cross is also correct, scoring 2 marks. The final mark is for stating that there is a 25% chance of a thin-furred offspring (two of the eight offspring in the Punnett square have the alleles 'tt' so have thin fur).

Question 6 **Variation and evolution**

(a) The graph below shows variation in body mass in a population of birds. What type of variation is shown in the graph? Give a reason for your answer. (2 marks)

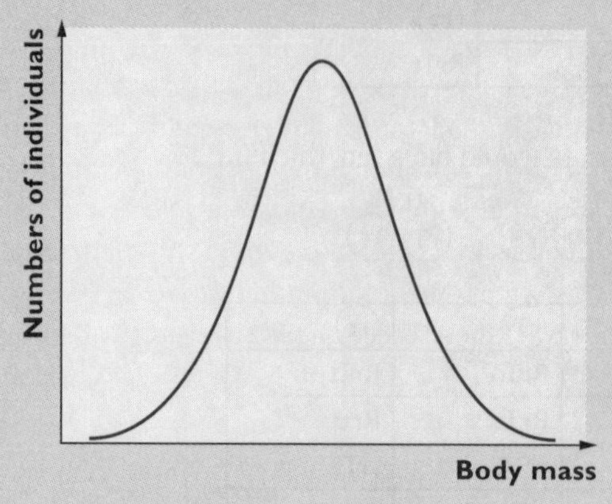

(b) A small group of birds left the main population and travelled to an island off the coast of the mainland. By chance, some of the allele frequencies in the gene pool of these birds were different from those in the gene pool of the main population. What name is given to this effect? (1 mark)

(c) Describe a mechanism by which the two groups could become different species. (5 marks)

e This question starts with straightforward application of key terms and then has a 5-mark section. It is important in questions with high mark allocations to make sure that you write in as much detail as possible using the appropriate key words and terms.

Student A

(a) The graph shows continuous variation because the line is continuous.
(b) The founder effect
(c) There is variation in the two populations. As long as there is no interbreeding, then, over time, through natural selection they could become different species.

e **4/8 marks awarded** (a) Continuous variation is correct but 'the line is continuous' cannot score the mark for explanation. (b) This is correct and gains the mark. Genetic drift is also an acceptable answer. (c) The student scores 2 marks here, 1 mark for 'no interbreeding' and 1 mark for 'natural selection'. The remaining 3 marks are for naming the type of speciation occurring (allopatric), mentioning how genetic drift could also lead to speciation and explaining what would happen once the populations had become different species (unable to interbreed to produce fertile offspring).

Student B

(a) The graph shows continuous variation. A variable such as body mass is controlled by a number of genes and is also affected by the environment. Therefore the data are continuous and not categoric.

(b) Adaptive radiation

(c) If there is no interbreeding between the two populations then, over time, they could become different species. This is an example of allopatric speciation because there is a physical divide (the sea) stopping interbreeding. They could become different species due to natural selection if the environment on the island is different from that on the mainland and there are different selection pressures. If there are no different selection pressures, speciation could still occur by genetic drift. Over many generations, the two populations would become so genetically different that they would not be able to interbreed to produce fertile offspring.

e **7/8 marks awarded** (a) This is an excellent answer and scores both marks. (b) It is possible that the idea of birds moving to an island has confused the student into thinking that the example is the same as the Galapagos finches. Adaptive radiation involves a common ancestor evolving into several different species adapted to different ecological niches. The example in this question is the founder effect — by chance, the new population has different allele frequencies from the original population. (c) This is a good answer and scores all 5 marks.

Question 7 **Applications of reproduction and genetics**

Cystic fibrosis is an inherited genetic condition. Sticky mucus builds up in the airways, leading to inflammation and infection.

(a) Explain how a mutation in a person's DNA leads to the condition cystic fibrosis. (4 marks)

(b) How can gene therapy using liposomes as a vector be used to relieve the symptoms of cystic fibrosis in the respiratory system? (4 marks)

(c) Other than liposomes, what else can be used as a vector in gene therapy? (1 mark)

e Most of the marks in this question come from two high-mark part-questions. This shows the importance of full revision. If a student has not revised cystic fibrosis fully then it will be difficult to write in enough detail to obtain all the marks. In a question such as this, the marks are awarded for key words and terms, so make sure that you include as many as possible.

Student A

(a) Cystic fibrosis is caused by a mutation in the gene that codes for the CFTR protein. This leads to the protein not forming properly and, therefore, being non-functional.

(b) First the gene for CFTR is removed from a cell from a healthy person who doesn't have cystic fibrosis. The gene is cut out using restriction enzymes. The gene is then inserted into a liposome. It is then placed inside the sufferer's body where it will cure the cystic fibrosis.

(c) A virus

e **5/9 marks awarded** (a) The student gains 2 marks for stating that cystic fibrosis is caused by a mutation in the gene that codes for the CFTR protein and that this makes the protein non-functional. However, there is no explanation of the consequences of non-functional CFTR, so the remaining 2 marks are lost. (b) The student gains the first 2 marks for describing the extraction of the CFTR gene from a healthy person and the use of restriction enzymes to cut the gene. However, the use of an inhaler to introduce liposomes into the respiratory system is not mentioned, or that the new CFTR gene must enter the nucleus, be incorporated into the DNA, and be transcribed and translated to form the correct CFTR protein. Any two of these points would have scored the other 2 marks for this question. The student also states that this treatment is a 'cure' for cystic fibrosis, which is not true. It would be a temporary treatment that would have to be repeated. (c) This correct answer gains the mark.

Student B

(a) There is a mutation in the gene for CFTR. This means that the CFTR protein does not form properly. This leads to chloride ions not being transported out of epithelial cells. This means that the water potential of the mucus is not lowered, so water does not enter the mucus from the cell by osmosis. This causes the mucus to be thick and sticky.

(b) A healthy non-mutated CFTR gene is removed from a donor cell from a person who does not have cystic fibrosis. The gene is cut out using restriction enzymes. The gene is placed in a liposome and an inhaler is used to bring the liposome into contact with the epithelial cell. It fuses with the cell membrane and enters the cell. If the CFTR gene enters the nucleus and transcription and translation take place, then the correctly formed CFTR protein would be produced.

(c) Bacteria containing recombinant DNA plasmids

e **8/9 marks awarded** (a) This excellent answer contains all the key points and scores all 4 marks available. (b) This is another excellent, detailed answer that scores all 4 marks. (c) The student slips up here on a simple factual recall question. Viruses are used as a vector in gene therapy.

Question 8 **Energy and ecosystems**

A simple food chain is shown below:

(a) Explain why all the light energy that falls on a shrub is not converted into chemical energy. (3 marks)

(b) Explain why all the energy that is converted into organic compounds by the shrubs is not available to the zebra. (2 marks)

(c) 'A zebra is a more efficient converter of energy than a lion.' Do you agree with this statement? Explain your answer. (2 marks)

e The important thing to notice here is that the questions are all worth 2 or 3 marks. This means that it is important to develop your answers, not just make one point.

Student A

(a) Some of the light is the wrong wavelength so is not absorbed by the chloroplasts.
(b) Some of the organic compounds are used by the plants in respiration.
(c) No I don't agree. A zebra feeds on plant material, which is hard to digest (for example, it contains lots of cellulose), so a lot of the energy in the plant isn't taken in by a zebra but is released as undigested food in faeces. However, a lion feeds on meat, which is rich in protein. Protein is much easier to digest than plant material such as cellulose, so the lion absorbs more energy from its food and less is passed out in faeces.

e **4/ 7 marks awarded** (a) This answer is correct. However, it is only one point, so the student scores only 1 mark. Another two separate points are needed to earn the other 2 marks. (b) This answer is correct, but as it is only one point, the student gains only 1 mark. The second mark is for stating that some of the energy is incorporated into indigestible material, such as roots or bark. (c) This answer is explained well if not concisely and scores both available marks.

Student B

(a) Some of the light might be the wrong wavelength to be absorbed by the chloroplasts, some will pass through the leaf by transmission and some will be reflected.

(b) Some of the products are used in respiration; others will form inedible material such as bark or roots.

(c) No, I don't agree. As zebras are herbivores their diet contains a lot of cellulose, which is difficult to digest. This means that they don't absorb a lot of energy, so don't convert as much as a lion.

e 6/7 marks awarded (a) This is a good answer that scores all 3 marks. (b) This is a good answer and scores both marks. (c) This answer is correct. However, it only scores 1 mark because there is no explanation of a lion being a more efficient converter because of its easier to digest protein-rich diet. In questions that ask for a 'yes' or 'no' answer followed by an explanation, the marks are for the explanation.

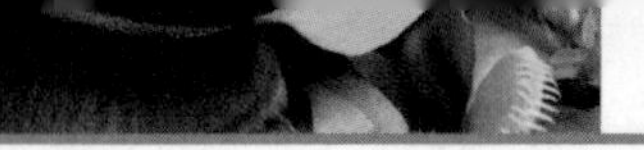

Question 9 **Effects of human activities and sustainability**

Describe the ecological consequences and possible solutions to:
- **deforestation**
- **overfishing**

(10 marks)

ⓔ This is an essay question that many students may feel confident in answering. However, they may not score many marks. The issue is that students often include unscientific information that cannot be awarded marks. It is important that you stick to concepts you have studied that are on the BY5 specification.

Student A

Deforestation is a major issue around the world. Lots of trees are cut down in many countries such as Indonesia and Brazil. Cutting down trees leads to an increase in carbon dioxide in the atmosphere, which is a contributory factor in climate change. It also leads to a large number of species becoming extinct, which is also a problem for humans. There are many ways to prevent deforestation. The key thing is to cut down fewer trees. This can be done with actions such as coppicing and long rotation times.

Overfishing is also a big problem. Many fish populations, for example cod, are collapsing due to overfishing. One way of minimising the effects of overfishing is to have nets with smaller mesh sizes. This means that younger fish are able to escape and then go on to reproduce. You can also have areas where fishing is banned for part of the year, particularly during times when fish are breeding. Fish farming can also minimise the impact of overfishing.

ⓔ **4/10 marks awarded**. This essay has too many general points that are not explained properly. By adding a little more detail (and not mixing up the point about mesh size), the student could have scored 8 out of 10 marks — much higher than the 4 marks achieved. This emphasises the importance of ensuring that your answers are explained fully and also of reading over your work carefully to eliminate silly mistakes. The student scores 1 mark for the carbon dioxide–climate change link and could have gained a second mark by mentioning that the rise in carbon dioxide is caused by a reduction in photosynthesis. Giving species extinction as a negative earns 1 mark. Just mentioning coppicing and long rotation times does not score — these must be explained. The student is confused about net mesh sizes. *Larger* mesh sizes are needed to allow younger fish to escape. The student gets 1 mark for discussing banning fishing in a particular area and 1 mark for stating that fish farming could also help reduce overfishing.

Student B

Deforestation is a big problem globally. One of the main environmental issues it leads to is an increase in the concentration of carbon dioxide in the atmosphere. This is because deforestation reduces the global amount of photosynthesis. As there is less photosynthesis, less carbon dioxide is fixed and, therefore, the atmospheric concentration increases. Another issue is the destruction of habitats, which reduces biodiversity. This is not only an issue for other organisms but also for humans. It is possible that plants that may have possible medical benefits for humans are becoming extinct before we are even able to discover them. Deforestation also increases soil erosion as the roots of the trees are unable to bind the soil. There are many ways to reduce the negative impacts of deforestation. Coppicing can be used. This involves cutting a tree down to the stump and then letting it regrow. This is a way of extracting wood without killing the tree. Also trees should be planted optimum distances apart. This increases the growth of trees and reduces disease and parasitism.

Overfishing is a major problem as fish populations are declining around the world. There are a number of ways of preventing overfishing. By ensuring that nets have large mesh sizes younger fish are able to escape and reproduce. There can be protected areas where fish are not taken. Fish farming can be used to produce fish without damaging natural stocks. However, there are many problems associated with fish farming, including an increase in disease. Quotas can be used to ensure that fishermen only catch a certain number of fish.

e **10/10 marks awarded.** This is an excellent, detailed answer that covers a whole range of both negative effects and possible solutions. By developing each idea fully the student has obtained as many marks as possible from each point. The student gets 2 marks for describing increased carbon dioxide leading to climate change due to reduced photosynthesis and 2 marks for explaining about loss in biodiversity and the possible loss of beneficial plants. Mentioning soil erosion as a negative scores 1 mark. The correct explanations of two methods of reducing the impact of deforestation — coppicing and planting trees optimum distances apart — score 2 further marks. The student has accumulated 7 marks, before starting to answer the second part of the question. This shows how you can build up marks quickly in essays by making detailed, well explained points. The student gets 1 mark for explaining the benefits of larger mesh sizes, 1 mark for protected areas, 2 marks for mentioning fish farming and its possible problems and 1 mark for quotas. Although this makes a total of 12 marks, the maximum mark that can be awarded in an essay question is 10. However, there may be as many as 15 possible marking points.

Knowledge check answers

1 Each DNA molecule consists of one original strand and one newly synthesised strand.
2 A gene
3 Nucleus
4 RNA polymerase
5 Through a nuclear pore
6 Ribosomes
7 Anticodon
8 Anaphase I
9 Chiasmata
10 Haploid cell
11 Seminiferous tubules
12 To nourish and protect the spermatozoa from the immune system
13 A polar body
14 In the oviducts (Fallopian tubes)
15 Capacitation
16 The formation of a fertilisation membrane
17 hCG
18 The anther
19 The transfer of pollen from the anther of one flower to the stigma of a flower of the same species
20 Cross-pollination
21 The ovule
22 The shoot
23 The physical characteristics of an organism as determined by the genotype
24 9:3:3:1
25 Crossing over
26 A mutation is a change in the volume, arrangement or structure in the DNA of an organism
27 Down syndrome
28 An agent that increases the risk of a mutation that may cause cancer
29 Not all DNA mutations are inherited. To be inherited the mutation must occur in gametes. Somatic cell mutations (those in ordinary body cells) are not passed on to the offspring and so cannot give rise to heritable variation.
30 The allele frequency will decrease.
31 Beaks
32 Interspecific competition is competition between organisms of different species. Intraspecific competition is competition between organisms of the same species.
33 Organisms that can interbreed to produce fertile offspring
34 Allopatric speciation
35 Homologous chromosomes are unable to pair up during meiosis I so gametes cannot form.
36 Genetically identical organisms
37 An undifferentiated cell that is able to divide to form specialised cells
38 Replacement of faulty genes with normal genes
39 Somatic cell therapy is altering genes in body cells; germ-line therapy is altering genes in gametes to prevent them from producing an organism with a genetic condition.
40 Chloride ions
41 Unpaired bases at the ends of part of a DNA molecule
42 Reverse transcriptase
43 DNA ligase
44 So that it is possible to determine which bacteria have taken up and are expressing the genes in the recombinant plasmid
45 Soya beans and tomatoes
46 A section of DNA that codes for proteins
47 95°C
48 Abiotic
49 Sunlight
50 GPP is the rate at which producers form products. NPP is GPP minus the products used in respiration by the plant.
51 Large differences in size between organisms in a food chain
52 $kJ m^{-2} year^{-1}$
53 A stage in succession
54 When warfarin is being used as a rat poison
55 The planned preservation of wildlife
56 Deforestation leads to a reduction in photosynthesis. This means that less carbon dioxide is fixed, so the carbon dioxide concentration in the atmosphere increases. This leads to climate change.
57 Larger meshes in nets allow younger fish to escape and go on to breed.
58 Fuels derived from plants
59 Biochemical oxygen demand

H

I

L

M

N

O

P

R

S

Modernising social services

Promoting independence
Improving protection
Raising standards

Presented to Parliament by the
Secretary of State for Health
by Command of Her Majesty

November 1998

Cm 4169

£14.50